The modern Cuban Government warranty designed and issued in January, 1931. See page 169 for the previous seal.

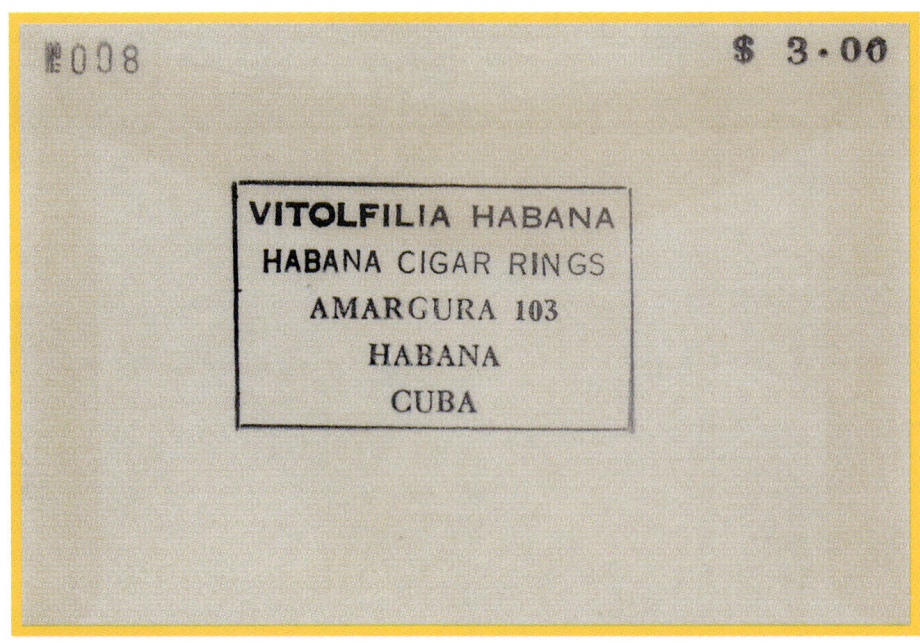

The collection of cigars bands, or rings as they are often called, is an international hobby. Many rings are extremely expensive. Ordinary, but out of print, rings are sold in Havana by Vitofilia on Amargura 103. The usual price is $3.00 per ring. On the facing page is a sample of Punch rings from a collector's book. Many collectors loaned the author pages of rings and labels with the strict agreement that they not be removed from the pages, repositioned or disturbed in any way. Thus some of the illustrations of these rings may be awkward.

# HAVANA CIGARS (1817-1960)

**by ENZO A. INFANTE URIVAZO**

Translated by Oscar de los Reyes and Carlos Lopez Cruz

© Copyright 1997 by T.F.H.Publications, Inc.
in cooperation with Habanos S.A.
Published by T.F.H. Publications, Inc.,
TFH Plaza, Neptune City, N.J. USA 07753
Printed, typeset and bound by T.F.H. in Neptune City, N.J.

# ABOUT THE AUTHOR

Enzo A. Infante Urivazo was born in Las Tunas, Cuba, in 1930. He is a teacher with an M. A. in political sciences, a Cuban History researcher, a freelance journalist, and a lifelong cigar enthusiast. His articles and essays have appeared in numerous Cuban journals and periodicals, and he currently is researching industrial aspects of the Cuban traditional cigar factories.

Since 1959, Urivazo has lived in Havana with his wife, three children, and four grandsons.

# ACKNOWLEDGEMENTS

The author wishes to express his gratitude to the following persons, agencies and institutions for their selfless support and collaboration which, to a considerable extent, permitted the necessary research, investigation and study to accomplish this book.

To Nancy Barreras, Mara Duharte and Mercedes Rodriguez from the State Archives in Pinar del Rio province; Berarda Salabarria, Carmen Gomez and Julio Lopez from the National Archives of Cuba; Martha Iglesias from the Party's Provincial Commission of History in Pinar del Rio; Noemi Dominguez Ramos, Zoila Artigas Calero and Mariluz Hernandez from the parent commission in the municipality of San Juan y Martinez; Enma Perez Martin from the Civilian Register in that municipality; Patricia Diaz, Otilia Morales and Teresa Campos from the Museum of the Revolution; Martin Duharte, Amparo Hernandez and Martha Cruz from the Cuban Institute of History; Obdulia Castillo, Martha Machado, Alicia Llarena and Zoila Lapique from the "Jose Marti" National Library; Maite Hernandez and Mercedes de la Fuente Garcia from the Tobacco Museum; Orlando Arteaga from the Association of Vitolfilians; Francisco Linares and Eduardo Gonzalez from Cubatabaco; Miguel Bellido de Luna from the "La Corona" factory; Jorge Concepcion, Maida Lorenzo, Crisanto Cardenas, Martha Cifuentes, Magali Garcia, Alfredo Malleza, Marianela Vena and Lazaro Garcia from "Partagas;" Joaquin Gomez from "H. Upmann;" Emilia M. Tamayo and Antonio Cruzata from "El Laguito."

Among those who generously and patiently gave me their time, wisdom, counsel and assistance in lengthy interviews are Dr. Julio Le Riverand, German E. Upmann Machin, Eduardo Rivera, Delia Jimenez Avellar, Graciela Tassende, Augusto R. Martinez Sanchez, Humberto Cabeza and Harold Beaton.

To Renaldo Infante Urivazo for his efficient collaboration in researching, drafting and checking all the material; to Mabel Oliva for her crisp and enthusiastic stenography; to Oscar de los Reyes and Aramis Aguiar for their sharp critical observations; to Cinda Pombo Bedia, Frank Carbonell, Angel Gomez, Jose and Maria Sanchez, Lianne Esquerra and my sons Enzo, Rosario and Pilar for their support.

To Herbert R. Axelrod and Bernard Duke for their constant solidarity and encouragement, and for their final editing of the English-language edition of this book.

# CONTENTS

**Preface** .................................................................................................. 11
**Chapter I-** BACKGROUND AND ORIGINS OF THE INDUSTRY ................... 17
**Chapter II-** EVOLUTION OF THE INDUSTRY AND COMMERCE ................. 25
**Chapter III-** THE BRANDS ........................................................................ 31
**Chapter IV-** THE FIRST WAR FOR INDEPENDENCE AND THE HAVANA CIGAR INDUSTRY ..... 33
**Chapter V-** THE CIGAR INDUSTRY AND CUBAN SOCIETY OF THE 19TH CENTURY ............. 37
**Chapter VI-** CHANGES AT THE TURN OF THE CENTURY .......................... 41
**Chapter VII-** BRITISH AND AMERICANS ................................................... 51
**Chapter VIII-** THE INDEPENDENTS ........................................................... 55
**Chapter IX-** INDEPENDENTS VS. TRUSTS ............................................... 59
**Chapter X-** MECHANIZATION OF THE INDUSTRY .................................... 69
**Chapter XI-** THE INDUSTRY SINCE THE THIRTIES ................................... 77
**Chapter XII-** EXPORTS DURING THE FIFTIES ........................................... 95
**Chapter XIII-** FACTORIES AND BRANDS .................................................. 96
**Chapter XIV-** HOW A HAVANA CIGAR IS MADE ...................................... 158
**INDEX** .................................................................................................... 171
**BIBLIOGRAPHY** ...................................................................................... 172

Cuba has historically honored the tobacco workers and tobacco farmers with the issuance of postage stamps. These are but a few of the stamps.

# PREFACE

The name **habano**, used to designate cigars made in Cuba, was not chosen by the Cubans. It emerged somewhat spontaneously in the international tobacco markets when certain features of undisputed quality led consumers to regard as best the products that came from Havana or were exported out of the Havana harbor. Once acknowledged as the best, intrigued consumers and traders began to wonder about the characteristics of the cigars, the peculiar traits that set them apart from the rest, their origin and their agro-industrial process. Some contended that the secret was in the leaves, others said it all depended on the hands of the cigar makers, still others argued that the merit had to be found in the grower's skill, in the selection, in the process of fermentation.

Through centuries of toil and progress, Cuban tobacco has been the object of numerous studies and its successive stages from the plantation to the factory have been thoroughly researched.

Nevertheless, whenever someone has come up with a supposedly exact, handy, universal and definitive formula, every attempt to reproduce the features of Cuban tobacco elsewhere has failed and exceptional quality tobacco continues to be found only in Cuba.

The factors that have contributed to the prestige of Cuban tobacco are found first and foremost in the nature of its soils. There are five clearly established tobacco zones or districts in Cuba where the plants grow remarkably well under the loving care of the **vegueros**. Vuelta Abajo, Semi Vuelta, Partido, Remedios and Oriente. Vuelta Abajo is the tobacco zone par excellence, covering fertile portions of the westernmost province of Pinar del Rio that include the sub-regions of El Llano, Lomas, Remates de Guane, the North Coast and the South Coast. Semi Vuelta lies lengthwise at the center of the province, from Heradura to Las Martinas.

The zone of Partido, in the province of Havana and part of the easternmost territory of Pinar del Rio, comprises the localities of Tumbadero, Caimito, Cayo La Rosa, Govea, Paletas, Piedras, Buenaventura and Artemisa amongst other cities.

Remedios is in the province of Villa Clara, but the tobacco with this denomination also includes the plantations of the neighboring provinces of Cienfuegos, Sancti Spiritus, Ciego de Avila and Camaguey, with important contributions from municipalities like Cabaiguan, Lomas, Manicaragua, Placetas, Santa Clara, Encrucijada, Camajuani, Vueltas, Yaguajay, Esperanza, Jicotea, Ranchuelo and Tamarindo.

The zone of Oriente includes the territories of Yara, Guisa and Jiguani in the province of Granma, Sagua de Tanamo and Mayari in Holguin and Alto Songo in Santiago de Cuba.

After the harvest, tobacco is classified according to its origin and its future use in the industry: Vuelta Abajo, the most valuable; Vuelta Arriba and Semi Vuelta, from the central and part of the western zones of the country; and Partido, whose name is derived from the characteristics of the leaves earmarked for filler or for being mixed as part of the blends that each factory makes. Tobacco classified as Partido can come from any zone in the country.

Endless hardships befell the tobacco growers from the very moment the leaves became a profitable merchandise. Going back into Cuban history, tobacco growers were among the first to rebel against the colonial regime.

The industry has travelled along a no less difficult and troubled road. The commercial trade of Cuban tobacco, be it in leaf, powder, rolled, pipe or cigarette forms, constitutes a cycle characterized by a succession of favorable and unfavorable junctions from the political, economic and social standpoints in which the prized plant has been a leading protagonist or, at the very least, a front row spectator.

The long list of events, alternating good times and bad times, begins in the early days when the colonial monopoly imposed its unfair rules. This was followed by the industrial awakening after 1817 with the emergence of small individual or family run workshops. By the mid 19th Century, an industrial boom occurred and, with the use of free and slave labor production and exports soared to record levels. Powerful captains of industry dominated the scene, while apprentices and Chinese coolies were enslaved at the factories. Fierce international competition generated protectionism and tariff barriers on the part of nations that tried to defend their own cigar industry against the Cuban exports.

Foreign cigar manufacturers purchased raw material from Cuba to undercut Havanas in a war of prices. Fakes and forgeries of Cuban cigar brands invaded the international markets, and they still do!

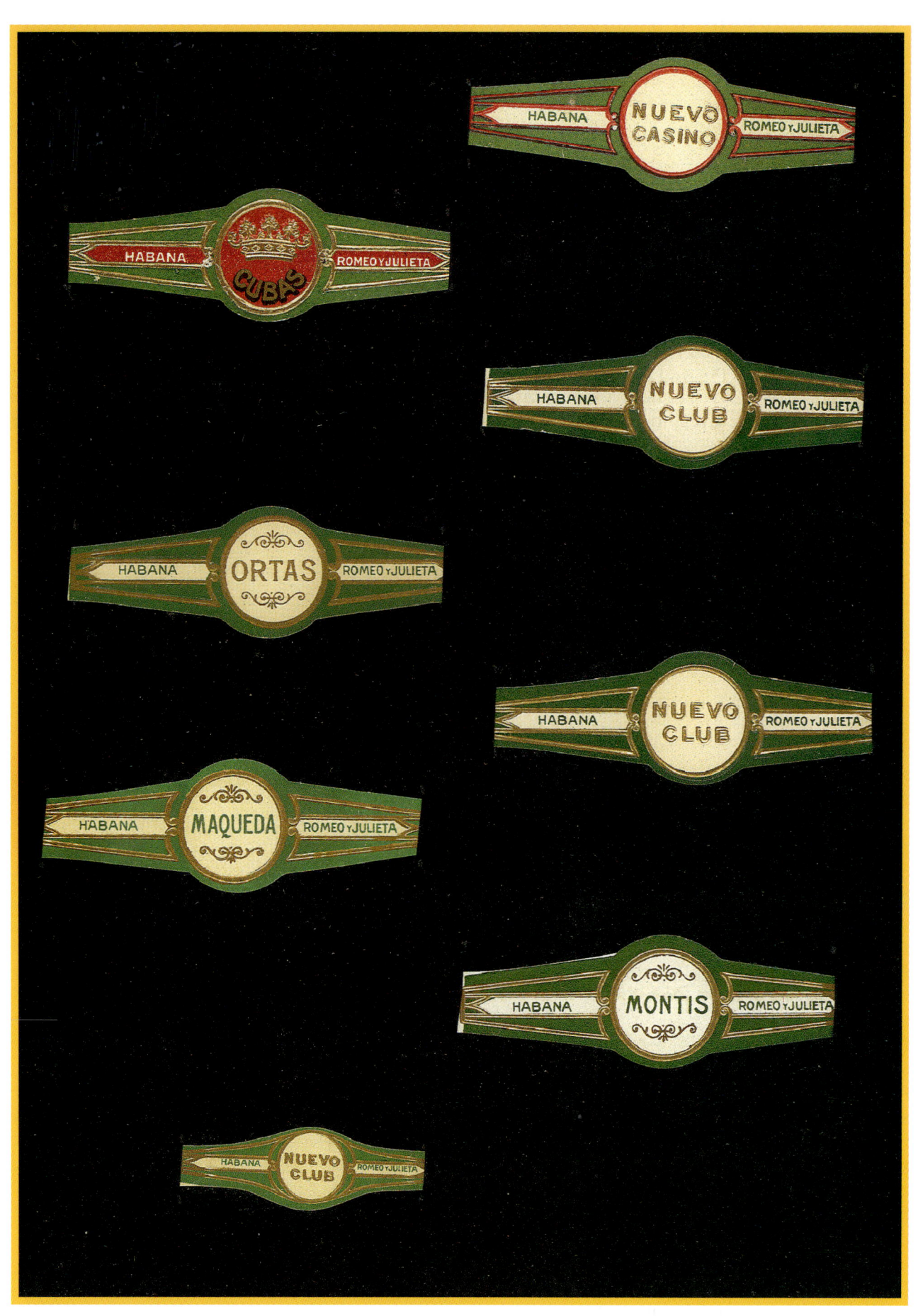

A sample of the vanity rings printed in Havana for Romeo y Julieta. Most manufacturers offer standard rings into which they can imprint the customer's name with orders of 500 cigars are more. The same is true today. You can buy rings with your name printed on them but the cigar brand name is missing. See examples of this on page 125. (Rings enlarged about 15%).

Production and exports plummeted during the wars of independence in the 19th Century and the two world wars in the present Century. The industry is hit hard by the economic crises of 1920 and 1929 and foreign investors put pressure on the Cuban producers.

The years between the two world wars are strife with clashes between workers and patrons over wage cuts, layoffs, shutdowns, attempts of mechanization and the transfer of nationally based factories to other countries in order to continue producing Havana cigars at lower costs.

A highly significant detail in the story of the Cuban tobacco industry was the absence of any effective official support, in contrast with the assistance received by manufacturers in other countries. That cruel orphanhood led the Cuban entrepreneurs to unite and organize in defense of their own interests in 1884. And it was precisely that lack of interest and neglect that paved the way for the swift take-over of the troubled yet powerful national industry by foreigners.

In less than a decade, over the turn of the Century, the transactions that radically altered the structure and organization of the Cuban tobacco industry were completed. The incoming British and American interests took almost absolute control, except for the obstinate endurance of the so-called "independent" manufacturers, some of whom would finally fall into the hands of the trusts.

Not until 1927 were any official institutions created to bolster the national industry. That year the **Comision Nacional de Propaganda y Defensa del Tabaco Habano** was set up and its constant activity in support of the industry and of the preservation and extension of credits for Cuban tobacco abroad, achieved a measure of success.

The very close ties between agriculture and industry is where the secrets of quality are found; this is what the first tobacco traders in Cuba perceived when they clearly realized that whoever controlled the entire process, from the plantation to the factory, would hold the key to sustained quality. Consequently, a Havana cigar is the exquisite expression that synthetizes those basic tenets. It is precisely their credit, prestige and excellence that has allowed the Cuban industry to survive and overcome such towering difficulties in the course of centuries.

A Havana cigar is the creation of the industry, the final product of an age-long art that is exactly reproduced time and again at the cigar maker's bench for the pleasure of smokers the world over. It is both an art and a science.

The legitimacy and unique supremacy of Havanas has been repeatedly and universally recognized. The Secretary-Treasurer of the Federal Commission of Commerce of the United States, Edmund A. Whittier, in a 1932 case against a manufacturer who unlawfully used the term **habano** on cigars that had not been made in Cuba, had this to say:

*"Tobacco harvested in the Island of Cuba has adopted the name **Habano** since the days of Columbus, undoubtedly deriving that name from the city of Havana, where it was first manufactured into pure cigars and from where the former and the latter have been exported under the respective denominations of 'Tabaco Habano' and 'Cigarros Puros.'*

*Due to its characteristic aroma and delicate taste, Cuban tobacco has been regarded thereafter as the best in the world for the elaboration of pure cigars and, consequently, pure cigars made with such an exquisite tobacco enjoy a fitting reputation as a unique product among smokers and traders. This reputation has existed for more than 300 years without ever being surpassed or even equalled by any other tobacco."* (*)

It is to the story of the industrial process, to the work of the craftsmen who impress their distinctive touch to each cigar, to the producers who raised the factories to levels of splendor, to the expert leaf selectors, traders, exporters, warehouse keepers, general workers and to the brands that have lasted to this day, that the present book is dedicated. It is not a detailed chronological history of tobacco; the author has tried to present smokers, scholars and the mere curious alike with a concise, suggestive outline of the makers of Havana cigars and their most famous factories.

In the following pages, interested readers will come into contact with the successes and failures, virtues and miseries that are inseparable from any human endeavor. Nevertheless, it tries to highlight the positive aspects behind every creative effort of man, which in this case resulted in a product that is attractive, enticing and universally consumed.

To please the tastes of millions by offering a very special pleasure in each vitola, is the highest aspiration of a cigar manufacturer. It is also the purpose of this tobacco producing island in the Caribbean, sole owner of its destiny and its industry, that with each cigar sends a genuine message of identity to smokers throughout the world: **The Habano.**

(*) In **Memoria de la Comision Nacional de Propaganda y Defensa del Tabaco Habano**. Report for 1932.

## XX.

## Oppidum Secota.

*OPPIDA palis non clausa plerumque amœniora sunt clausis, vt ex hac pictura patet, quæ oppidum S E C O T A ad viuum exprimit: hinc inde enim sunt sparsa ædificia, & hortos, E. in quibus nascitur Tabaco, ab ipsis Vppowoc appellatum: syluas item, in quibus ceruos capiunt, atque agros, in quibus suum frumentum serunt, continent. In agris tabulatum exstruunt supra quod ædiculam seu tugurium hemicycli in modum tectum, F. in quo vigil esse possit: nam tot sunt volucres & feræ, vt, nisi diligenter excubias agerent, breui omne frumentum absumendum sit, eam ob causam vigili sine intermissione vociferandum, & strepitus faciendus. Serie quadam H. illud serunt, alioqui alia stirps aliam extingueret, nec frumentum debitam maturitatem G. consequeretur: nam eius folia ampla, magnarum arundinum foliis similia. Aream etiam habent peculiarem, C. in quam cum suis vicinis conueniunt ad solenne illud suum festum celebrandum, vt XVIII. pictura demonstrat: deinde locum, D. vbi peracto festo conuiuaniur. Ex opposito rotundam habent aream: B. vbi ad preces solennes faciendas congregantur: à qua non procul distat amplum ædificium A. in quo Regum seu Principum tumuli, vt ex vigesima secunda figura apparebit. Hortos etiam habent, quibus melopepones alere solent, littera I. notatos, locum similiter, K. in quo ignem solennibus festis struunt, & paulo extra oppidum, flumen, L. vnde aquam hauriunt. Isti ergo ab omni auaritia liberi, genialiter dunt axat viuunt: sed eorum festa noctu celebrantur; eam ob causam luculentos ignes struunt, ne in tenebris versentur, & vt suam lætitiam testentur.*

# 20
# Secota Village

The villages are not enclosed by stakes (fences) for the most part and are delightful to look at, as is revealed by this illustration, which displays a living, working Secota village: for on this side and that the buildings are few and the gardens, **E**, in which tobacco is produced, form what is named VPPOWOC itself: and so these fields, in which they capture deer, and other fields in which they produce their grain: the villages contain all these parts.

In the fields they heap up a story above which a small temple or cottage is built in the manner of a semicircle, **F**, in which field they keep watch, for there are so many wild birds that unless they diligently keep watch, all the grain which is short lived will be consumed, for which reason the vigil is maintained without interruption as they cry out with a loud noise.

In a certain row, **4**, they plant this and in others stems of plants are overturned and grain which ought to be mature is tilled into the soil, **G:** for the leaves are full with similar plants ready to be harvested. Indeed they have a private area, **C**, in which they assemble with their neighbors for the purpose of celebrating their annual feast. Then from this place, **D**, they are brought together for this particular feast. They have a round place on the opposite side, **B**, where they gather to make solemn prayers: from which place at a distance there an ample building in which the kings and leaders gather, from whom on the following twentieth day a figure will appear.

Indeed they have gardens in which they offer little petitions, as noted in letter **I**, a similar place, **K**, in which they build up a fire for the solemn feasts, a little outside the village, a river, **L**, from which they draw their water. Therefore, the children also live jovially from this observance, but they celebrate this feast at night; for which reason they build up brilliant fires, so that it is not observed in the shadows in order that they might assert their happiness.

**This illustration and description appeared in a book dated 1593. The title of the book was faded but it appeared as** *Admiranda Narratio, Fida Tamen de Commodoset.* **The author acknowledges Dr. Mark Zilberqvit for his generosity in allowing me to copy the pages from his very valuable and delicate book.**

**Not all Havana cigars are made in Havana! Cigars made in Santa Clara, Cuba are also considered as Havanas by lovers of Cuban cigars. (Enlarged 50%).**

# Chapter One
# BACKGROUND AND ORIGINS OF THE INDUSTRY

The year 1817 was an important one for the history of the tobacco and cigar industry in Cuba because it was then that the abolition of the monopoly was declared in the island. Rigorous control of production had always been the general rule of the colonial administration, which purchased the harvest, determined the volumes to be exported to Spain and fixed the prices, usually unfair to the hard-working growers.

During the 18th century, the uprisings of the tobacco growers in the vicinity of Havana were a clear indication of the fact that there was an explosive situation in the agricultural sector, which was not at all happy with the exploitation to which it was subjected. The areas around Jesus del Monte, today a busy urban thoroughfare, witnessed the hanging of 12 tobacco growers charged with heading a rebellion against the mandatory and abusive terms imposed by the colonial government.

Those events, along with the experience deriving from the conditions established by the English during their short occupation of Havana in the year 1762, gave the island's residents a new perspective of the tobacco trade and instilled in them a much more deeply-rooted rejection of the monopoly. The freedoms given by the English in the commercialization of tobacco brought great profits to the growers. The increasing demand for Cuban cigars was spreading and their fame was far flung. With the return of Havana to the Spanish crown, however, the monopoly was quickly reestablished. In 1764 a royal decree again imposed the system of strict control and gave the hated factory absolute control under the former regulations, which had been in place since the establishment of the monopoly in 1716. There followed 54 years of unrest and hardship during which the vices and corruption of an insensitive and opportunistic bureaucracy fanned the flames of hatred for the monopoly. They were years of calamities which a chronicler of the times described as a period of "errors and horrors carried to such an extreme that there was hardly a single official of that agency (the factory) who was not brought to trial."

The decision to abolish the monopoly came like a fresh breeze and it was a wise step, despite the limitations it introduced. There is no doubt that the economic and political moment in Spain was a determining factor in the decision, but there were other factors as well, such as the state of opinion of the illustrated sector of the colonial society. Francisco de Arango y Parreno was one of the most noteworthy and influential Cubans of the times. He was Counsellor of State and held several important positions in the colonial government. His studies and essays on agriculture, commerce, communications and industry, contained in 19 works that he left for posterity, gave him great authority in those areas and made him a mandatory reference for government decisions. Early in the 19th century, the Superintendent and

**Don Francisco de Arango y Parreno.**

General Director of Tobaccos of the Island of Cuba asked his advice in connection with a possible price increase on leaf tobacco. The illustrious son of Havana took the occasion to go much further and give him a comprehensive overview of the entire tobacco system. In 1805 he wrote a rigorous and well-substantiated report in which he defended the fairness and effectiveness of liberalizing the growing, manufacture and commercialization of tobacco.

His appeal did not produce any immediate effect, but it did channel a certain current of thought –something which only a person of his stature and prestige could have done –which drew supporters and in the space of a few years rallied them around his ideas. In 1811, the Titular de Estado de la Real Hacienda (colonial Minister of Finances) raised his voice in the **Cortes** using Arango's arguments to demand the abolition of the monopoly and freedom of trade.

Ferdinand VII was reigning in Spain, having been reinstated after the Napoleonic nightmare of the beginning of the 19th century.

His so-called "illustrated despotism" changed the face of classic and anachronistic absolutism. New methods and forms were vying for preeminence in the politics of the nations of Europe, occasionally compelled by the contradictions and pressures of the international correlation of powers and forces. It was this monarch then, moved by the circumstances, who on June 23, 1817, decreed the abolition of the monopoly and did away with the privileges of the factories. The step was not really complete and did not do away with the injustices, for it imposed exaggerated taxes on tobacco growers and traders providing that "the twentieth part of the harvest shall be the royal tribute to be paid."(1)

The end of the monopoly and liberation from the clutches of the factory should have

**Fifty years ago (1946) huge factories with hundreds of cigar rollers were established.**

When the trade in Cuban cigars dropped, rollers were discharged and the factories had about 30% of their normal staff. Only the most skilled and productive workers were retained.

produced the immediate emergence of the cigar industry in Cuba and its fast and progressive growth, but the iron-fisted taxation policy was a stumbling block which made that impossible. However, a few small shops did appear as early and groping pioneers. For more than a decade they developed and increased until by 1827 they had acquired a certain significance.

During the course of the preceding years, the metropolitan administration had been letting up and introducing successive cuts, culminating that year with the total suspension of all taxes on tobacco farming, manufacturing and internal consumption and fixed only export taxes at a rate of $1.00 per quintal (equal to 100 pounds) of leaf tobacco "if it was exported on ships flying the Spanish flag; $1.75 on ships flying foreign flags; $0.75 per thousand cigars, whichever the flag of the exporting ship."(2)

As of that year, there was a take-off of the cigar industry with much more powerful companies. Private initiatives of the kind that moved small cigar makers began to draw investors. Exporters and other traders started to refaccionar offer growers and cigar makers premiums to ensure their supply of the products to participate in a market that was becoming increasingly attractive and competitive. Some of those which later developed into big factories and important cigar exporters got started precisely in this year. Some thirteen

King Ferdinand of Spain in 1818.

**Vanity rings issued by Romeo y Julieta about 1932. (Same size).**

years after the abolition of the monopoly by Ferdinand VII, the industry was thriving as a result of the immense efforts of the pioneer founders who devoted their energies to their factories, as well as to the opening of new foreign markets, leading to a solid and progressive consolidation of the industry.

In 1827, ten years after the year the monopoly was abolished, cigar exports amounted to 407,000 units; by 1836, the figure had leapt to 4,887,000. In Havana alone, there were 306 cigar factories with 2,152 workers.

The decade that began in 1840 brought an era of splendor to the tobacco industry, raising it to the rank of second most important contributor to the national income.

The Real Sociedad Economica de Amigos del Pais (Royal Economic Society of Friends of the Country), an institution founded late in the 18th century during the administration of Don Luis de las Casas, played a key role in the development and consolidation of the cigar industry at the time. The Sociedad Economica attracted the most outstanding personalities in Cuban society. The capacity, knowledge, zeal, determination and expertise of its members were put at the service of the country and, ever since it was founded, it had a hand in every positive step taken for the advancement of the island, as well as in all endeavors that entailed development and progress. It played a support and advisory role to the colonial administrations, but invariably addressed its energies toward the benefit of the island and maintained its own independent positions in the most diverse and controversial matters. The Real Sociedad designed projects, executed plans and provided an opportunity for the participation of many illustrious natives of Cuba and of Spain. It enjoyed immense prestige and fostered the application of science in agriculture and industry, supporting the initiatives of outstanding personalities in culture, science and business. In connection with the cigar industry, aside from the many studies it contributed, the Sociedad Economica played the role of official controller to guarantee brands and organizational methods in the cigar agroindustry and trade; it also organized the first exhibits of manufacturers and introduced the specialty of cigar rolling in its School of Apprentices. Of a total of 853 enrollees in 1840, 178 were in cigar manufacturing. Over the next two years, the number of apprentices aspiring to become cigar rollers increased sharply to 268 in 1841 and 314 in 1842.

One of the most important elements in connection with the development of the industry was the new class of merchants, whose warehouses –huge, overflowing buildings, sometimes silent, other times humming with activity– spread throughout the different neighborhoods of Havana, although most were near the factories or grouped in certain areas to which they contributed their own outward characteristics and bustle, generally close to the docks or to the railroad depots. These warehousers operated not only as buyers and exporters but quite often financed the advancement of the industry with risk investments in cigar rolling shops and in new cigar brands.

The industrial boom and thriving markets also entailed dangers.

In many factories where cigars were rolled, only men were employed. This soon changed however and in modern times (1990's) there are more women than men.

The old monopoly had never been able to do away with the contraband trade and according to creditable sources, actually encouraged it. Under the new conditions, smuggling continued to prosper since the forces in charge of repressing it did little and the corrupt bureaucracy in power was getting rich off the profits of the underground trade. But other evils were even more damaging. A plethora of parasites developed as a result of the disorderly growth of the industry, booming markets, new mercantile relations, increasing production, commercial voracity and the fame and acceptance of the Havana cigar. As far back as 1837, responsible voices were alerting to the dangers and calling for measures. There was a fear that growing ambitions in connection with the trade and unscrupulous dealers would damage the prestige of the Havana cigar and undermine its exceptional quality. The main danger came from foreign operators whose only interest was turning a fast profit with no consideration at all for the consolidation of the industry's potential and the future possibilities of what had already become part of the country's natural wealth.(3)

Fakes and counterfeits were turning up everywhere, infecting the markets, eating away at the Havana's paradigmatic quality and making it the target of every hustler in the business. But the foreigners were not alone. The disease also spread inside the country, to the point that in 1842, the Captain General of the island issued an energetic decree forbidding the use of all brands other than those approved by the Civil Government and warning engravers to abstain from printing seals for manufacturers who did not have that approval.(4)

The counterfeiters inside and outside of Cuba were trying to supplant the older, more

prestigious brands that enjoyed the preference of the consumers. In the 40s, there were already a number of patrician ancestries and quality credentials in which the consumers placed their trust. One of them was Bernardino Rencurrel, founded in 1810, before the abolition of the monopoly, probably thanks to a royal dispensation for being a purveyor to the Crown. H. de Cabanas, also established since 1810 under the protection of a similar royal favor, had actually been in the trade since 1797, although there are no documents confirming that date. Jose Garcia's Mi Fama por el Orbe Vuela (My Fame Circles the Globe), one of the longest and most pompous of brands, was founded in 1830, and La Lealtad, created by Jose Morejon y Roja, was founded in 1831 on a street in Havana which came to be known by the name of the factory and is still known as Lealtad Street today.

Por Larranaga, one of the cigar brands with the greatest impact on the international market, was born in 1834 El Figaro by Julian Rivera. Punch and El Sol, by Marcelino Borges, were founded in 1840, and in 1844, another brand that acquired world fame, Partagas, by Jaime Partagas, was officially founded, although some date its origins as far back as 1827. In 1844, one of the first non-Spanish foreigners, certainly the first German, set up shop with the H. Upmann brand. That same year La Reforma de Jose Ruizs´ La Africana saw the light, and was soon assimilated by the well-known firm of Pino Villaamil & Compania.

In 1845, the brand destined to become one of the most famous in the world, La Corona, was created by Jose de Cabargas and taken over in 1882 by Alvarez, Lopez & Co., under whose management it gained its greatness. That same year witnessed the birth of El Huracan and La Meridiana, both amid an avalanche of less noteworthy firms.

**Calixto Lopez y Ca. produced a line of cigars called *Los Reyes de España*, literally *The Kings of Spain*. These were very special cigars. (Same size).**

1-Gonzalez del Valle, Angel: Memorandum to the **Comision Nacional de Propaganda y Defensa del Tabaco Habano.** Siglo XX Printing Press. Havana, 1929. p. 18.
2-Ibid, p. 19.
3-Ibid, p. 21.
4-Ibid, p. 22.

Vanity bands from H.Upmann were very common because of the fine name of *Upmann*.

Vanity rings produced by Partagas.

## Chapter Two

# EVOLUTION OF THE CIGAR INDUSTRY AND COMMERCE

During the 40s, production and consumption figures still indicated that the market was dominated by cigars manufactured in Cuba with the coveted and palatable leaf grown in the region known as Vuelta Abajo. There was, however, a great deal of speculation about what was genuine and what was false among the many products circulating on the market with those appellations. Between 1826 and 1830, the industry exported more than 245,000 pounds and between 1835 and 1840, the export volumes surpassed 790,200. In 1850, production of Havana cigars amounted to 1,275,540 pounds and nine years later, in 1859, the figure increased by an additional 481,178 pounds. This very brief analysis of two decades indicates that cigar exports increased by 123% and although for obvious reasons there was a more pressing demand for leaf tobacco, the profits from the cigar trade were ultimately quite significant and compensatory.

On an initiative sponsored by the Real Sociedad Economica de Amigos del Pais, described above, in 1847, Havana hosted the First Public Exhibit of Products of the Industry, which awarded a number of medals to local manufactures for their products. Mr. Julian de Rivas won a gold-plated silver medal for the quality, variety and presentation of his cigars and another silver medal was awarded to Mr. Jose del Rosario Jimenez, for his exhibit of rolled cigars.

There can be no doubt that Cuban tobacco, in any of the forms in which it was sold, was enjoying a moment of supreme international predominance. The Royal Society of Industry and Commerce of Havana complained at the time of the lack of reliable statistics and, although it did use the General Balance, published under the auspices of the Real Hacienda (the colonial ministry of finances), it had no other alternative than to resort to certain speculative estimates in calculating the projections that reflected social and economic relations within the industry. Thus, to calculate domestic consumption, it assumed that the white population of the island in mid-century was 420,000 and that the free colored and slave population amounted to 470,000. It estimated that there were some 280,000 whites and about 120,000 "colored" who smoked cigars every day. It then estimated that those 400,000 smokers consumed 3 cigars a day, for a total of 1,200,000 a day and a total for the year of 438 million.

In 1847, a worker could hand-roll about 300 cigars a day, on the average, for which he was paid a salary of 1 peso 4 reales.

Supposing that the average worker was able to get in 20 days of work a month, his total production for the year would amount to 72,000 cigars. Considering cigar export statistics—close to 200 million—, total production would amount to some 600,000 a year, which, taking as a basis for calculation the individual 72,000 output, would lead us to infer that there were over 8,300 workers employed by the industry, which paid out wages of 3 to 3 and a half million **pesos fuertes** (strong pesos) a year.

**The factory of J. Suarez Murias in Havana. These old factories were heavily protected with iron gratings over the windows which were padlocked to protect the contents of the factory from strikes and thefts.**

The intention behind the Royal Economic Society's calculations was to highlight the economic importance of the industry, which produced great profits for its owners but also employed a great many workers and made a substantial and steady contribution to the public coffers. The arguments were, in fact, advanced by the cigar industry barons, who were increasingly demanding a reduction in the taxes levied on them.

The entire taxation structure governing exports had changed since the taxes were established back in 1827 and in the 1850s the state of opinion was that the burden was particularly heavy on rolled cigars. The manufacturers wanted to take some of the burden off the cigar itself and shift some of it on to the leaf so as to free their hands to deal with foreign competition and strengthen the Cuban product in international markets, which were feeling the pressure of manufacturers in England and France, who were importing Cuban tobacco and manufacturing cigars in their own factories. After 1850, these concerns gained support and advised the adoption of protective measures.

Despite these and other threats, the cigar industry was firmly consolidated in Cuba and Havana was the hub of activity. Factories were springing up everywhere with their own brands.

After the ups and downs of previous decades, the cigar industry had entered into an investment frenzy and a wild rush had been unleashed to get a piece of the business. The most successful factories and shops soon abandoned their old and often modest enclaves and moved into impressive premises built in the classical, neoclassical and renaissance styles. The old city, with its urbanistic conception based on the somber and sober colonial, romanesque and late gothic styles, was suddenly studded with buildings of impressive beauty in which we can still admire the classical facades, the arches and the Dorian columns framing the well-ventilated and comfortable, high-ceilinged halls, where the manufacturing process took place, along side the administration and management annexes.

In 1859, the H. de Cabanas y Carbajal factory turned in statistics that illustrate the growing importance of those manufacturing firms in Havana. Whereas their total investment for the year 1852 had been 60,000 strong pesos, seven years later they were investing 120,000. During the period they had produced 41.3 million cigars, an average of 5.9 million a year, and had a payroll of 300 workers and employees of all kinds.

Statistics made public in 1852 revealed that during the decade that ended in 1859, cigar exports had totalled 246,863,000 units, an impressive figure which was, nevertheless, 4.44 million fewer than in 1854.

There were at the time, in and around Havana, 516 cigar shops and factories owned by 377 cigar and cigarette manufacturers, for some had more than one manufacturing facility. These establishments employed 15,128 workers, who supported 45,384 dependents and family members with their income.

The total number of cigar manufacturing shops throughout the country was 1,733. All together they produced the 246,863,000 cigars mentioned above as exports for the year 1859 and an additional 374,285,000 for the domestic market, for a grand total of 621,148,000 cigars.

Despite the cigar industry boom, there were times of crisis. In 1856 the local press reported that there were some 5,000 cigar makers out of jobs in Havana, a depressing situation that was attributed to extremely high exports of leaf tobacco, which reduced the supply available for internal processing and left the industry without its raw material. But the leaf exporters were not exclusively to blame. As it happened, some time earlier, rumors had begun to circulate about a possible increase in import tariffs to the United States. Thus, in the year before 1855, cigar exports set an all-time record with sales of 356,582,500 units. And sure enough, on March 3, 1957, the Congress of the United States passed a new tariff of 30% ad valorem on cigars to take effect as of July of that year.

The rush preceding the duty hike set off a chain reaction that induced importers to stock up, thereby cutting heavily into the imports of the following year and leading to production drops and, of course, layoffs.

As of the drop in exports in 1859, a progressive decline set in as a result of increasing competition from factories abroad and the pernicious effects of the contraband trade.

In their desperation over the problems of the crisis and the loss of markets, the Cuban manufacturers turned to the Colonial power with the old and hackneyed issue of the tobacco monopoly, which had been retained in Spain and continued to be the target of constant virulent attacks. Since almost all of the manufacturers in Cuba were of Spanish origin, when anything happened that went against their interests, such as the measure adopted by the Government of the United States, they turned to the Motherland with petitions, supplications and

Vanity rings from Partagas.

**Benito Celorio was president of the Cigar Manufacturers' Guild in 1879.**

demands for a liberalization of trade with Spain so as to have access to that export market. In 1865, the Real Sociedad Economica de Amigos del Pais got behind the demands and added its own weight and authority to the balance of contradictions and debates by calling on the Crown to abolish the tobacco monopoly in Spain for the good of the nation.

In 1877, three of the most important men in the development of the cigar industry, Julian Alvarez, Prudencio Rabell and Antonio Rivero, delivered their demands to the central Government. In 1879, Benito Celorio, president of the Cigar Manufacturers' Guild, and Vicente Galarza, president of the Agricultural and Industrial Center for the Tobacco Industry, added their weight, as captains of the tobacco agroindustry, to the demands. In a letter addressed to the Captain General, they said, that they "would be obliged to show greater gratitude by a measure that reestablished free sales of cigars in the Motherland." (1)

The two appeals sent by the industrialists and by the Real Sociedad were obviously moved by a single purpose, but they were far apart in time and in their presentation of the facts. Spain had, in fact, suspended the monopoly and decreed freedom of trade on a test basis in 1876 as a result of pressures from within and from the colonies. But the test did not last long. Hardly a year had gone by when everything reverted to the former iron-fisted monopoly control. The Crown alleged that smuggling had increased, revenue had dropped and corruption was rampant in all sectors.

This is why the Cuban industrialists referred to a "re-establishment" of free trade. The defenders of free trade, were angered and offended by the reversion and they denounced the pressures of opposing interests bent on the failure of a system of relations that was beneficial for both parties. They complained of the absence of complementary instruments and measures that should have been implemented in support of the abolition. The Minister of Colonial Affairs at the time, who was opposed to the liberalization, said that free trade "was the sore through which the wound hemorrhaged." (2)

Despite the official intransigence, aspirations were dictated by the situational and quite uncertain outcome of the strife between those who, in Cuba and Spain, insisted on the need for change and those who tenaciously clung to the monopoly, confining tobacco to the role of revenue element. During the last 20 years of

**Don Saturnino Martinez, secretary of the Chamber of Commerce, Industry and Navigation in Havana in 1892.**

Don Segundo Alvarez was president of the Asturian Center. He presided in 1901-1902 and was reelected for 1903. He was president of the Chamber of Commerce, Industry and Navigation in 1892.

Very large bands or rings were produced with magnificent artwork (usually produced in Germany) and were used by such grand old cigar brands as Murias.

the century, in 1877, a law was passed in an attempt to respond to the most recent demands of the Cuban industrialists. The so-called Tobacco Monopoly Lease Law put all cigar imports in the hands of a Leasing Company, which was obliged to purchase 3 million kilos of leaf tobacco from Cuba per year. Leopoldo Carvajal, Marquess of Pinar del Rio, was appointed to represent the company in Cuba. Ensuing events revealed that the decree was continuously violated.

In their "Exposition on Tobacco," delivered to the Ministry of Colonial affairs on February 9, 1892, D. Segundo Alvarez and Saturnino Martinez, President and Secretary, respectively, of the Chamber of Commerce, Industry and Navigation in Havana, concluded the appeal with the following statement, "...in which case requests a modification authorizing unrestricted sales of that product in the Peninsula compensating with the necessary taxes for any loss this measure might entail for the revenues of the State; as expressed by the corporations on different occasions, as one of the ways to prevent the disappearance from this island of the cigar industry, which is a very principal source of its wealth and prosperity." (3)

1. Gonzalez del Valle, Angel: op. cit. p.24.
2. Celorio Hano, Benito: Letter to His Excellency the Governor General of the Island of Cuba, Havana, January 27, 1879. In the National Archives, ledger 159, single page.
3. Gonzalez del Valle, Angel: op.cit. p.27.

Some of the vanity bands produced by Romeo y Julieta, La Corona and others, were fairly sophisticated. There were usually lithographed in six colors, gold stamped and embossed. These skills have been modernized and the stone lithography used to produce these old bands has long ago been lost when four color lithography came into being at the turn of the 20th century (1900-1910). (Enlarged 15%).

**Chapter Three**

# THE BRANDS

During the years before 1856, cigar brands were registered with the Ordenanzas Municipales (Municipal Ordinances). To get a brand entered into the registry required presentation of the corresponding authorization from the Civilian Governor. We have seen above that a number of measures were implemented to try to prevent falsifications. For their part, lithographers and printers operating in Havana kept careful records of the factories and firms for whom they made seals or hierros, cliches? As early as 1848, to preclude all doubts about the propriety of its relations with its clients, one Havana printer presented to the authorities a book containing 232 "views" or "sketches" of the brands for which it made lithographs in its shop.

This was the way the printers responded to the municipal authorities, who were unhappy, not with the frauds and forgeries, but with what they considered an arbitrary commercial practice by the different firms, which were continuously bringing out new brands. The problem became such a nuisance in the meetings of the town council that on February 23, 1859, the City Government of Havana resolved to request of the Civilian Governor that the different cigar companies should have a single brand, even if it meant changing the municipal ordinances, because of the abusive practice of many manufacturers of registering an immense number of brands for themselves.

The writ filed to this effect spent a few months getting through the official levels until March of 1860, when the request was denied despite the protest of the City Government.

The circulars issued by the Civilian Government in 1856 and 1870 had the force of legal precepts governing brands. Twelve years later, in 1882, the rules in force in Spain were implemented in Cuba, and four years later, new standards were dictated, which were again amended between 1900 and 1929.

Before 1887 there were 348 cigar brands registered in the country.

It was estimated that the most important factories made between 12 and 15 vitolas (as the different cigar types and sizes are known) and that in 1859, Havana's 377 manufacturers were using some 800 brands. (*)

The following brands were registered in the period following 1850:

La Carroza de Venus, El Carro de Febo, La Artemisa, La Dominica,

Las Tres Gracias, Las Tres Coronas, La Rosarito, El Tio Canillita and La Pilarsito all appeared in 1853, a year of notable growth in the cigar industry. El Veguero appeared in 1854; La Explotacion, by Carlos Medina, in 1858; La Charanga, El Moro Muza, by Diego Gonzalez, and El Rico Habano, by B. Menendez y Hermano, in 1860.

**A magnificent old band from Eden by Bances y Lopez in Havana. (Same size).**

Vanity rings from Calixto Lopez in Havana.

## Chapter Four
# THE FIRST WAR FOR INDEPENDENCE AND THE HAVANA CIGAR INDUSTRY

On October 10, 1868, Carlos Manuel de Cespedes, a rich lawyer and plantation owner of the region of Bayamo and Manzanillo, who was destined to go down in Cuban history as the "Father of his Country," headed an uprising in the locality of Yara, in Oriente Province, near the Sierra Maestra mountains. The eastern provinces were not important tobacco producers as were the central and western provinces, where there were important plantations in Vuelta Abajo, Partido and Vuelta Arriba.

No one ever seriously believed that the confrontation would affect the tobacco plantations or the cigar factories, even if it spread to the western part of the country. The most significant tobacco areas were in Pinar del Rio, Havana and Villa Clara and, because of the importance of the industry for the capital city, it came under the protection and influence of Spanish military might.

However, the political contradictions and the fact that the war was for independence did have an impact on every aspect of the life of the nation, including the tobacco industry. A number of historians specializing in that period have said that after the beginning of the first war for independence, the manufacture of Havana cigars split in two: those made in Havana and those made in the United States, particularly in Tampa, Key West and New York.

The cigar industry had been established in the US for quite some time, but the new twist was that now there were Cuban manufacturers with Cuban cigar makers using Cuban leaf. The move north had been motivated by political contingencies or by the new facilities open to them at a time when in their own country conditions were abnormal and the difficulties were growing. Many of these expatriate manufacturers were successful, but there are two names that should head the list: Vicente Martinez Ibor, born in Valencia, Spain; and Eduardo Gato, born in Havana.

In a personal account offered years after his exile, Martinez Ibor described the lightning journey that took him from Cuba to Tampa.

He had arrived in Cuba years earlier in the employ of the Spanish Government. He soon got into the cigar manufacturing business and, in 1859, his factory, El Principe de Gales, was already considered important. He never got involved in politics, neither for nor against the independence struggles, and devoted all his energies to his factories, minding his own business, to coin a phrase. He had signed a contract with the administration of the Havana prison whereby the inmates made cigars for his company.

This increased his production capacity, something he very much needed because the orders were pouring in and his shops could not cope with them. In 1871, he was shocked to learn from a friend, Vicente Galarza, that an order had been issued for his arrest.

**As the business of making cigars left Havana for Tampa, Florida to escape U.S. prohibitive taxes, fancy cigar bands were produced to cover their use either in Tampa or Havana since they were always printed in Havana. This is a huge band. (Same size).**

Galarza, who was himself above suspicion from the Government, hid him in his home. All the contacts made by his friends to free him of the arrest order failed and he had no other alternative but to try to flee the island. He left Cuba aboard a schooner to which he was taken, personally and in his own carriage, by the man who had hidden him, Count Galarza. Galarza's home was later ransacked by a mob of **Volunteers**.

Martinez Ibor was known as an intelligent and hard-working man who was thoroughly familiar with the cigar business. He established his residence in Tampa and, taking advantage of the facilities he was given, set up his business and factories there, in a part of the city which came to be known as Ibor City. His example spread like wild fire and soon other Spaniards and Cubans, from investors and businessmen to cigar rollers, fled the situation in Cuba and established themselves in Tampa, Key West and New York. Men like Salvador Rodriguez, Isidro Pendas, Haya, Lozano and many others opened factories to manufacture Havana cigars.

Among the advantages offered in the United States to those who undertook the move were: a reduction in customs duties, the right to use the appellation "Havana" for their brands, and the protection of the Government, as opposed to the lack of official support the industry had to put up with in Cuba. The transfer of a sizable part of the industry to the United States was a hard blow for the product made in Cuba. In spite of it all, however, the dedication of the Captains of Industry and the prestige of the Havana cigar in the international marketplace prevailed and in 1879, the year the war ended, cigar exports amounted to 182,355,720, of which 56,066,325 were sold in the United States.

With the advent of peace, manufactures in Havana turned their energies against the counterfeiters, who were now very strong in several states of the Union. In 1877, Julian Alvarez, Prudencio Rabell and Antonio Rivero made public demands for the enforcement of international law against the counterfeiters operating in the neighboring country.

**Vanity bands from Jose Gener.**

Vanity bands from Romeo y Julieta.

An assortment of rings from El Rey del Mundo.

Chapter Five

# THE CIGAR INDUSTRY AND CUBAN SOCIETY OF THE 19TH CENTURY

In 1865, at a time when the situation of cigar workers was particularly stable, the free cigar workers organized the Mutual Aid Society of Craftsmen of Havana, the first such organization in Cuba at the time. The goal of the Society was to aid its members, to prepare to face difficulties together in time of crisis and to promote the education of its members. The Fraternal Association of Santiago de las Vegas and the Society of Craftsmen of San Antonio de los Baños were founded soon afterward, both in localities with weight in the cigar agro-industry.

That same year, a new craft was born in Cuba, the cigar-shop reader, an institution of historical importance that became a tradition and has been preserved to the present. The cigar-shop reader was provided with a privileged position in the rooms where the cigar rollers or other craftsmen worked and a lectern from which he would read, in a loud and powerful voice, selected works of world literature and the local press. The practice flourished and became a daily activity because the manufacturers soon realized that the rote manual activity of the craftsmen was fully compatible with the new intellectual activity and could constitute a symbiosis capable of increasing cohesion, rhythm and labor productivity.

Julian Rivero's El Figaro factory was the first one where the voice of the reader was heard. It was followed some months later by the Partagas factory, which instituted the reader on January 9, 1866. History has it that the rostrum or lectern was an initiative of Don Jaime Partagas himself, who donated the first one for his workers.

This new possibility of complementing manual labor with intellectual education was a breakthrough that was applauded by all for its particularly humanitarian aspects, but for the Cuban people it was of special historical importance because some of the forefathers of the Cuban nation turned the lecterns into tribunes from which to spread the gospel of Cuban independence, both in Cuba and in the United States, including the most outstanding of them all, the man who was to become the Apostle of Independence, Jose Marti.

**Julian Rivero's *El Figaro* factory was the first factory to employ readers.**

On August 24, 1880, the Manufacturers' Guild was founded. Its first provisional president was Mr. Juan A. Bances, who was replaced by the first elected president, Antonio Allones. In October of 1884, the Union Society of Cigar Manufacturers of Havana was formed by the merger of the Manufacturers' Guild and the Association of Cigar Manufacturers of Partido. The ceremony was presided over by Don Leopoldo Carvajal, Marquess of Pinar del Rio, who was subsequently elected its first president, seconded by Lucio Arena as Vice-President and Gonzalez Alvarez as Secretary. This association worked along lines similar to those of the Economic Society for the protection of brands and the defense of the industry as a whole, as provided for in a resolution issued on January 30, 1888, by the Ministry of Colonial Affairs.

On February 13, 1889, the Union of Cigar Manufacturers of Havana was authorized to allow its members to place on their cigar boxes a printed seal as a guarantee of origin. The rectangular seal had on its left side, under the royal crown, the coats of arms of Spain and of the City of Havana; on the right side it had an engraved bust of Christopher Columbus and in the center it read: "The Union of Cigar Manufacturers guarantees the origin and legitimacy of the cigars bearing this seal and will, pursuant to the law, prosecute anyone attempting to falsify or amend it."

**Dr. Don Juan Bances Conde founded the Association of Cigarette Manufacturers which later merged to form the Union of Cigar and Cigarette Manufacturers of the Island of Cuba. Bances became the Presidente General.**

The measure and its enforcement were good, but they were not able to fully achieve their objective. The falsification of Havana cigars continued both inside and outside of the country, but the counterfeiters now had to reproduce the guarantee seal as well.

Later, in January of 1896, the Union of Cigar Manufacturers of Havana merged with the Association of Cigarette Manufacturers (founded by Mr. J.A. Bances) to form the Union of Cigar and Cigarette Manufacturers of the Island of Cuba.

Up until 1879, work in cigar factories was limited exclusively to men. Men held all the jobs in the factories and shops and no one could even imagine the presence of women in the places where such a unique craft was practiced. As of that year, however, things changed and women were given access to the cutting, stemming and rolling tables. The first factory to take that step forward in history and open up its premises to women was La Africana, owned by Pino Villaamil and Co. In due time, the other factories were to follow suit and women workers became one of the reasons which nurtured the legend of Havana cigars and helped develop their mystique. There are, among the features that characterize the Havana cigar, certain steps, "touches" and abilities in the manufacturing process in which the presence of the "craftswoman" has played a key role in the prestige and quality achieved.

Before 1890, relations between workers and owners developed without significant discrepancies. The cigar workers had organized to defend their interests and increase their education, and the fact that they were a key element in an industry of no small economic import for the nation gave them a certain air of respect.

The manufacturers, for their part, were well aware of the specialized nature of cigar-making operations and were out to get the best craftsmen and hold on to them. For decades, the main problems affecting the workers had to do with crises of the industry, market fluctuations, taxation and political differences between the colonial power and the colony. Nevertheless, the records do mention the first cigar-makers strike, which took place in 1866 at the H. de Cabanas y Carvajal factory. Labor leader Saturnino Martinez and sixteen of his colleagues were arrested with the intention of expelling them from the country, but they were defended by a prominent personality and lawyer, Jose Ignacio Rodriguez, Secretary of the Economic Society of Friends of the Country, who won their acquittal.

It might seem extraordinary for the Secretary of such a staid and proper institution to appear in public in a case such as this one, but the characteristics of its members were such that the organization always projected an image of independence. The Royal Society had, on several occasions in the past, disagreed with the authorities and defended its opinions and attitudes firmly. It was not by chance that, in 1868, with the outbreak of the war for independence, it was censured by the colonial authorities for closing down in a veiled gesture of protest against the colonial power for having been deaf and blind to its warnings.

There were other strike movements in 1886 and 1887 that were considered important. They were headed by a controversial character by the name of Jose M. Aguirre, born in Asturias, Spain, who made cigars for a number of factories: La Flor de Cuba, Partagas, La Corona and La

Escepcion, up to 1889. A few years later, in 1893, he founded a newspaper, *El Tabaco*, and from then on shifted his loyalties to the side of the manufacturers.

In all the time before 1890, labor demands centered on industry-related issues such as higher salaries, shorter working hours, better and more humane working conditions in the shops and factories, and job security, always threatened by the specter of layoffs.

After the middle of the century, the idea of the cigar-rolling machine began to occupy some minds. One anonymous author, according to the records of the Economic Society, wrote about the use of cigar-making machines; cigarette machines were already in use in the industry.

**His Excellency Sr. D. Leopoldo Carvajal, the Marquis of Pinar del Rio.**

**Rings from the firm of Maria Guerrero.**

**Rings did not always have the same shape and sizes and a great deal of variety in shape, size and color existed for many years during the ferocious battles between cigar competitors at the turn of the 20th century.**

## Chapter Six

# CHANGES AT THE END OF THE CENTURY

During the Ten Years War (1868-1878) for Cuba's independence, the Spanish Government tried to shift the cost of its military operations and the contingencies of a war situation on to Cuban export products. The first step came the year after the war began, in 1869, with a tax of two silver escudos/1,000 cigars exported and 25% on the industrial quota.

In 1890, the Budget Law of the Island for the fiscal year 1890-1891, put a 20% surtax on imports, which raised the duty on flour and edible fats imported from the United Stated to 50%. The successive surcharges levied by Spain on these items led to significant reductions in imports. Flour imports dropped from 343,000 sacks and barrels to about 40,000.

The reaction in the United States did not take long. William McKinley, who at the time was a Senator from the state of Ohio, introduced into the US Congress a bill to amend the customs structure, which ultimately came to be known as the McKinley Bill.

For all practical intents and purposes, the bill excluded Cuban sugar and tobacco from the US market. Trade relations between Cuba and the United States had prospered in the preceding decades in terms that were beneficial for both sides. Cuba was a sure market for edible fats, flour and the output of US light industry.

Conversely, the United States was a stable and important market for Cuban raw materials and agro-industrial products.

In the tobacco industry, we must recall that in 1857 the US applied new duties against leaf tobacco and cigars, and raised the duties on cigars considerably in 1883. A number of localities in the United States already had important cigar industries based on Cuban leaf and the additional duties were intended to blunt or eliminate the competition from cigars manufactured in Cuba.

However, despite these events, consumer preferences determined a sustained presence of the genuine Havana cigar on the US market.

The McKinley Bill set off a veritable chain reaction. In Cuba, all the commercial and industrial organizations mobilized for action, the most energetic being: the Chamber of Commerce, the League of Businessmen, the Circle of Plantation Owners, the Economic Society of Friends of the Country and the Union of Cigar and Cigarette Manufacturers, led by an ad hoc committee called the Economic Propaganda Committee. The general agitation they unleashed and the overall state of public opinion came to be known as the Economic Movement which, despite its official acceptance did not escape the ill feelings and personal rejection of the Captain General, Camilo Polavieja.

Through their powerful organization, the cigar manufacturers had a hand in all the "hottest" moments of the campaign. The Spanish Government could not remain indifferent to the unrest and urged the group to elect a group of Commissioners to travel to Madrid to present their case. The Union of Manufacturers elected

**President William McKinley of the United States declared war with Spain in April, 1898. He was also the author of the McKinley bill which just about prevented Cuban tobacco products from being sold in the U.S. It was probably more effective than the Helms-Burton Bill of 1996, or President John F. Kennedy's embargo.**

Benito Celorio Hano, who was a lawyer and founder of the Association and had close ties with the pioneers of the industry. He was accompanied by Dr. Portuondo, representing Santiago de Cuba, Segundo Alvarez, representing the plantation owners, and Rafael Montoro, representing the Economic Society of Friends of the Country.

**Don Benito Celorio.**

Six meetings or conferences were held in Madrid on December 23-30, 1890, under the chairmanship of the Minister of Colonial Affairs, Antonio Maria Fabie.

Although he had no specific representation powers, Don Segundo Alvarez, who did have a very strong interest in the industry, put all his support behind the demands. He maintained that there were in the United States two cigar industries: one mixed, which used tobacco grown in the United States in a blend with tobacco imported from Cuba; and another industry that used Cuban tobacco exclusively and was almost completely owned by a colony of rich Spanish businessmen who had factories in Tampa, Key West, New York, Chicago, St. Louis, New Orleans and other cities. These manufacturers, who had been born in Spain, were competing against both the American-owned factories and the Cuban manufacturers.

Convinced that the McKinley Bill was the result of the influence those manufacturers had already developed in the Congress in their war against the genuine Havana cigar, Don Segundo Alvarez inveighed against critics maintaining that "they oppose us fiercely and, in their impassioned drive, they accuse us of lack of patriotism, and all because we are defending a Spanish industry in Spanish territory against a foreign industry in a foreign territory." (1)

Celorio had a more pragmatic approach. He called for concessions to the United States on flour and fats in order to get reciprocal benefits for Cuban sugar in the final agreement covering cigars, given the key role the United States market played for the Cuban cigar industry. While cigars were not included in the reciprocity agreement, the concessions would pave the way for further duty agreements and would help prevent the collapse of the Cuban industry. In an appeal that was as heroic as it was pathetic, he said, "The Cuban cigar industry is not asking for and does not need protection, but it needs consumer markets and all its quality will be to no avail if it is excluded from the markets, as has been happening. Let us make whatever concession we can in the present case so that in the future we will not have to lament our lack of foresight and justice in abandoning such a great number of persons to desperation." (2)

At the time of this crisis, the wealth represented by the cigar industry was estimated to be in the vicinity of 25 million pesos and, on the human side, considering the agricultural aspect alone, the industry supported some 60,000 persons. Annual tobacco production was, on the average, in the order of 430,000 quintales (one quintal=100 lbs.), of which 50% was consumed by the industry. About 100,000 quintales were processed into cigars: 260 million cigars with an estimated value of 12 million pesos and whose main market was the United States, which accounted for 118 million cigars a year.

The duties paid before the McKinley Bill amounted to 2 pesos and fifty cents, but the Bill increased the duties by 80% so that an additional 10 pesetas had to be paid over the original 12 and a half pesetas. Thus the exporters insisted that while the previous duties were all but prohibitive, the new ones were practically a ban.

Other world markets were not in a position to absorb the full Cuban output and all of them had regulations that bordered on protectionism. Following the US market, and in descending order of importance, came: England, with 42 million cigars a year; Germany, with 30 million; Central and South America, with 25 million; France, with 12 million; Spain, with 6.5; and Austria, Italy and the rest accounted for another 26 million. So the elimination of the most important market was, up to that moment, the cruelest and most deadly blow ever

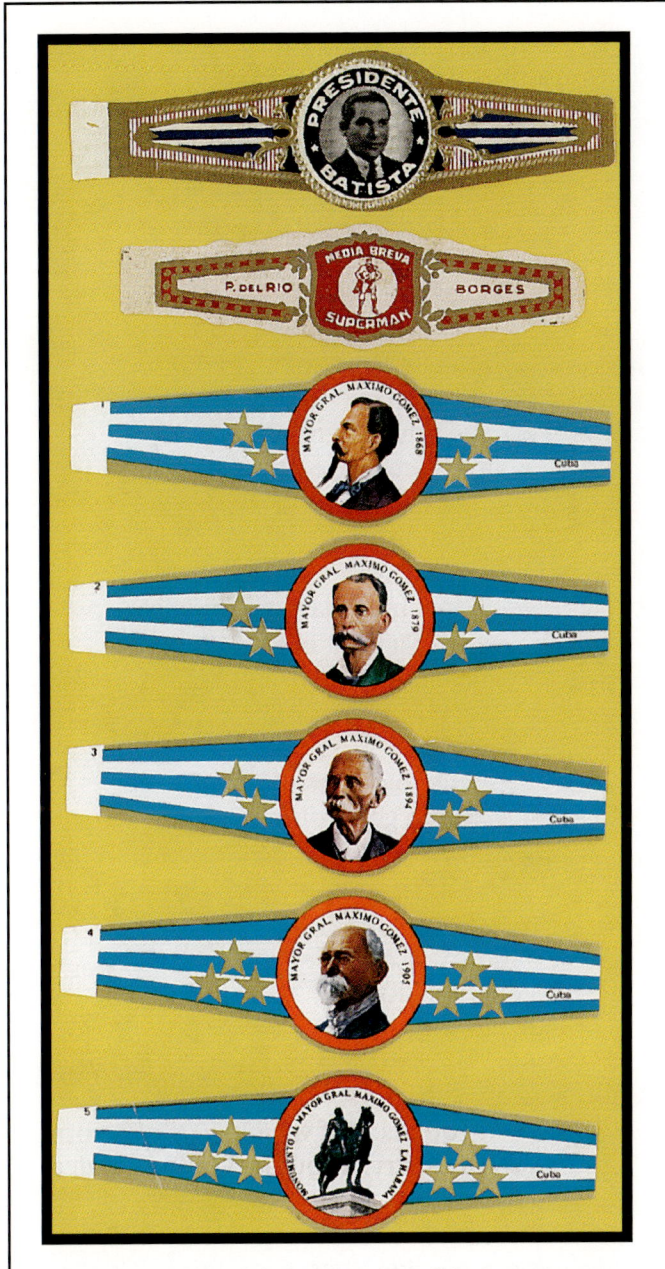

The Cubans have honored personalities, personages and heroes throughout their history. One of the most famous heroes in Cuba's war of independence against Spain was Maximo Gomez. Gomez was a great military leader who was born in the Dominican Republic but came to Cuba to help it gain its independence.

Gustavo Bock.

delivered against the Cuban cigar industry. Thus a collapse became a real possibility and with the bankruptcy of the industry, other markets would fold in its wake. At that time, the US produced cigars exclusively for domestic consumption, but if the Cuban industry collapsed leaving market vacuums around the world, no one could rule out an attempt to fill those vacuums with American cigars. "Our cigars are the best in the world," cried the manufacturers bitterly, "but the most ill treated."

As a result of the six long and heated conferences in Madrid, on January 4, 1891, the Commissioners made ten proposals, including the following, in connection with the cigar industry:

4th- That, without prejudice to the agreement or correlation to which the preceding paragraph refers, the signing of an agreement with the United States should be promoted in the fastest and most effective manner possible so that the duties in their new rates for Cuban tobacco will be reduced in consideration of the fact that 50% of the leaf and about 45% of the cigars are exported to that country..."

6th- That, given the aggressiveness of the new North American tariffs, which put tobacco from the island in the most precarious circumstances, and recalling that the market in the Peninsula is all but closed to it, help should be provided for this important source of wealth by suppressing the export duties without further delay." (3)

A memoir published in 1894 by the Union of Cigar and Cigarette Manufacturers of Havana contained certain opinions on the covert reasons behind the customs reform in the United States. "...In a nutshell," it said, "they want to pay us for the sugar and tobacco they consume with the articles they produce. This is the gist of Senator McKinley's amendment: for sugar the aforementioned bill recommends a hiatus to conclude next July 1st, but for tobacco, the

Various vanity bands.

44

reform is for such immediate, terrible and radical implementation that it is tantamount to locking us out of that market." (4)

The severe economic problem engendered by the Budget Law and the McKinley amendment was to mark the future of the cigar and tobacco industry in the ensuing decade and would bequeath the new century a somber panorama. The political, economic and social events leading up to the second revolutionary outbreak would seem to be reproducing situations already experienced in the period preceding the insurrectional conflicts of the 70s. On the former occasion, during the course of what was called the Junta de Informacion (Information Meeting) the Cubans were denied the concessions they were demanding. On the latter, following an equally formalistic process, the answer was the same: mockery and disappointment.

The Minister of Colonial Affairs promised to review the commitment of the Spanish Sociedad Arrendataria (Leasing Society) to purchase 3 million kilograms of leaf tobacco from Cuba, which, according to the Commissioners, was not being fulfilled, and the Government issued a final order that the Society was to receive the cigars made in Cuba on consignment.

As if fatal events came in droves, in 1891 Argentina also increased its customs duties on imported cigars.

The cigar manufacturers started to penetrate the colonial political parties, perhaps in the conviction that their interests should become of the public domain. They issued a manifesto addressed to the voters in the city of Havana, signed by Pedro Murias, Melchor Fernandez, Bernardo Martinez, Juan Valle, Antonio Fernandez Garcia, Francisco Menendez and Benito Suarez. This manifesto gave rise to heated polemics within the ranks of the Union Constitucional Party which split the party and produced a splinter group that was to become the Reformist Party, headed by cigarette manufacturer Prudencio Rabell.

The failure of the Commissioners in Madrid and the implementation of the customs reform in the United States were a hard blow to the Cuban cigar industry and many of its leaders were overcome by disappointment, doubt and fear. Consistent with the prevailing psychological and emotional state among Cuban manufacturers (so important in business decisions in times of crisis), a group of representatives of US manufacturers came to Havana and organized the transfer of factories to Florida, contracted workers and even lured a number of growers.

Although the United States said that the Cuban manufacturers were not receptive to the tempting proposals, it would seem that not all remained indifferent to the mermaid songs for many manufacturers, large and small, as well as workers allowed themselves to be seduced. The fact is that a locality with a cigar manufacturing tradition as strong as the one that existed in San Antonio de los Banos, bordering on Havana, where it was known that there were about 1,500 people making cigars, was suddenly left without manufacturers in mid-1891. By 1893, even the United States was admitting that some industry owners had traveled to Florida to install large manufacturing plants and that three of the most important manufacturers were involved in negotiations for a similar move. (5)

And it was not at all surprising. The offers became more and more interesting as the attractive elements were made more explicit.

The US Government was offering:

a. Construction of a building with a 60-foot front by 100 feet;

b. For all new factories, enough land for the construction of 200 homes;

c. A 100,000 peso credit with 7% interest on the amounts used;

d. Cashing of manufacturers' checks without bank commission.

The conditions required that the manufacturer turn out 10 million cigars in the first three years, or sooner if possible, following which he would receive the deeds as legitimate acquirer of the referred properties.

In commenting on these proposals the Cuban manufacturers underscored their impact on a dying industry and attempted to rally and organize efforts to avoid being absorbed. They revived the debate on the same theses rejected in Madrid and again demanded:

—That our industry be exempted from all taxes except the municipal tax;

—Establishment of trade treaties to facilitate markets for the products;

—In the introduction of any innovation into the trade treaty with the United States, to reject the exclusion of cigars;

—To open the market of the Peninsula to our products.

The effects and the decline continued inexorably and in 1893 many manufacturing licenses were returned producing a 20% drop in the output of the industry. The factory shut-down at that moment was also determined by a recurring measure. The Draft Budget for the coming fiscal year contained a plan by the Minister of

Colonial Affairs to further increase export duties. Thus the factory shut-down became a kind of protest.

Parallel to this, other events, such as the discrepancy between the Chamber of Commerce and the Union of Manufacturers, helped complicate the period even further. The crux of the contradiction was an attempt by the manufacturers to shift the entire burden of the duty increase on to the warehousers. The battle was staved off by the intervention of a very influential personality, the Count de la Montera, who prevented implementation of the measure. (6)

The year 1895 was to bring the convulsed political panorama and the complicated state of the economy, with its impact on the industry, into much clearer focus. That was the year when the Second War for Independence broke out in the small town of Baire, in Oriente Province, after long years of preparation abroad. The names of Jose Marti, Maximo Gomez and Antonio Maceo rallied increasing influence in the island but, as on other occasions, the western agro-industrial provinces trusted that the war would remain confined to the other extreme of the island without affecting their interests.

As one chronicler of the times has written, "On October 24, 1895, Antonio Maceo began to implement, at Mangos de Baragua, the plan for the invasion of the west and ninety days later, on January 22, 1896, to the surprise of strategists the world over, he presided over that memorable and solemn session of the Municipal Council of Mantua."

Mantua was Cuba's westernmost municipality, far in the west of Pinar del Rio, the province where the famous Vuelta Abajo tobacco plantations were located. Maceo and Gomez had led their troops from one end of the island to the other, destroying enemy armies and burning down the most important cash crops. It was in Mantua, practically at the doors to Havana, where he designed his famous Western Campaign.

The situation was changing fast for this part of the country and the panic-stricken cigar manufacturers rushed to Captain General Valeriano Weyler. The famous edict prohibiting exports of leaf tobacco grown in the provinces of Pinar del Rio and Havana, except as required by the needs of the Peninsula, was dictated on May 16, 1896.

The edict was intended to preserve the raw material so that the factories would not be forced to close. It allowed 10 working days for the discharge of previously acquired commitments, but it, too, was the object of violations and conspiracies by the officials charged with enforcing it. The Union of Manufacturers had serious run-ins with the colonial administration over the concession of shipping permits not provided for in the edict.

Some manufacturers tried to find their own solutions. Don Gustavo Bock, the former owner of El Aguila de Oro, who had entered into a merger in 1887 with Julian Alvarez's Henry Clay company to form a limited liability company chartered as Henry Clay and Bock and Co.

Ltd., which became one of the biggest companies in the tobacco and cigar business, reacted to the dangers posed by the war and its duration by deciding to find a place that would be safe from the torch and from combat action. He promoted the emergency development of large tobacco plantations on the Isle of Pines, off Havana's southern coast, which he continued to exploit until 1899.

The war was spiraling by the day and, as a result, all exports based on agriculture had to bear the brunt of the consequences.

The competition abroad was quick to take advantage of the situation and, turning deaf ears on the internal measures dictated by the colonial Government at the behest of the manufacturers, spread the news around the world that the famous cigars made with Vuelta Abajo tobacco no longer had the same quality because, as a result of the war, the industry had been obliged to use leaf of inferior quality.

The Union of Manufacturers decided to mount a vigorous counterattack and, in an open letter published in the London press, categorically denied the insidious rumors. The letter, dated in December of 1896, ended with a declaration by the manufacturers stating that they were firmly resolved to close down their factories temporarily before turning to subterfuges that would diminish the prestige of their brands.

The last decade of the 19th century brought with it events of such far-reaching importance that they were destined to change Cuban society forever in many ways. The cigar industry, built on the efforts of half a century of dedication, was also on the threshold of unprecedented changes. It was the result of hard work and constance which took it from its initial experimental phases to full maturity and made it a thriving and competitive industry thanks to the technical rigueur and expertise attained and maintained by its craftsmen and promoters.

General Antonio Maceo, Chief of the Cuban rebel forces in Pinar del Rio, 1895-1896.

Jose Marti, apostle of Cuban independence, was also a reader to the cigar workers in Tampa and Key West, Florida. In his readings he influenced many Cubans.

In order to offer something different, Upmann made vanity bands in order to honor people and associations without have the honorees pay for the privilege. The rings added prestige to the Upmann line and were considered endorsements.

47

**Hoyo de Monterrey featured wide bands, heavily embossed. (Same size).**

The country might have developed, like so many others, into a simple grower of tobacco and manufacturer of cigars. But in its lands, its plantations and shops, in the minds and hands of its growers and craftsmen, something wonderful happened that quickly distinguished the product as the finest in the world.

The pride of the country's industry and victim of whims, harassment and ambition. Even in the days of unbridled splendor, it had to navigate through complex labyrinths of supply and demand and suffer from hostile markets which, in their awareness of its supremacy, rejected the presence of its quality. They tried to deprive it of its birthright, to falsify it and to don its regalia: all in vain; and confronted by their failure, they tried to besmirch its prestige and discredit it.

At the end of the century, the Cuban economy impotently contemplated the agonistic transformation of the tobacco and cigar industry. Tobacco agriculture was emerging from the convulsive post-war situation and demanding the rehabilitation of its traditional cultivation areas. The established and accredited manufacturers, stunned by the confused political situation and the closed markets, were retreating, disarmed, before the opportunistic attack of foreign capitals anxious to swallow them up or replace them. Along with their plans they brought with them the new structures typical of capitalism in expansion, characterized by the formation of groups of brands and manufacturing facilities which required a standardization of the product. English and American investors assaulted the manufacturers and organized syndicates and trusts that quickly took over most of the cigar businesses, including firms with long-standing and recognized international prestige.

The recovery of tobacco agriculture brought to light the old mistakes and historical shortcomings. Everything that had to do with the seeds, cultivation, treatment, the factory process and orientations for the growers and dealers was always handled as something to be taken in stride as a function of practice and experience and soon solidified into a an empirical agro-industrial culture completely ignored by the country's authorities. Except for the work of a few Spanish and Cuban scientists and the concise instructions given to the growers by the factories, there was nothing else, for the colonial governments never supported the creation of study and research centers. Thence the importance of the old School of Apprentices founded by the Economic Society of Friends of the Country in the 1840s, which trained the craftsmen.

In the competing countries, the agronomy schools, experimental stations, specialized institutes, laboratories doing industrial research and scientists consecrated to the quest for new varieties and hybrids made up the technical and academic foundations on which to build a solid future. In Cuba all the experience and expertise was disperse and cached in individuals who were crushed by disappointment. Many of them were Spanish and reacted to the political changes by packing up and going home or by seeking new horizons in other lands.

And yet, the anxiously awaited and necessary recovery did take place in due time, spurred on by the experts who remained. Many of them still preserved their reserves of perfectly classified seeds, and when these seeds were not enough, they imported more from other places, particularly from Puerto Rico.

The fear that the quality of the Havana cigar would be lost was quickly dispelled as it became clear that its superior qualities came from the land, from the characteristics of the soil, from the blends and combinations of fertilizers, from the expertise in the treatment of the plants and from the climate, the salt-laden air itself which sweeps over the lands of Vuelta Abajo giving the leaf its particular gifts and helping to create that exquisite bouquet that can only come from a Havana cigar.

Marti said that the tobacco growers treated their plantations "as if each plant were a delicate woman."

The industry quickly developed into a two-branched structure: one made up by the big monopolies and another by staunch independent producers. The former had to deal with the contradiction of two counterposed interests which were soon conciliated through the consolidation of manufacturing facilities with their processes centralized in order to achieve the standardization of the product. The latter maintained that their traditional working methods were part of the essence of the quality of the Havana cigar. Both became very strong in international markets, with a certain competitive balance and more than a few arid and circumstantial polemics.

1-Speech by Segundo Alvarez, representative of the commercial interests of the Island, at the Fourth Conference of the Minister of Colonial Affairs with Cuban and Puerto Rican delegates to discuss the Law of Commercial Relations between the Peninsula and The Antilles, Madrid, December, 1890. Memoir of the **Union de Fabricantes de Tabacos de La Habana** containing the most relevant accomplishments of the corporation from 18 September, 1890 to 5 February, 1894 in defense of the general interests of the industry, Havana, 1984. p. 179.

2-Ibid, p. 188.

3-Ibid, p. 16 et seq.

4-In the Report on the Assessment of the New Tariff, unanimously adopted at the Joint Assembly on 5 May, 1890, addressed to H. S. The Minister for Colonial Affairs. pp. 13-15. Published in Havana, 1893.

5-Gonzalez del Valle, Angel: Op cit. p. 49.

6-Fernandez Roque, Mario: **Cuba, el Pais del Tabaco Habano**, in **Libro de Cuba**, Havana, 1953.

Upmann with their endorsed bands. The *Lord Lonsdale* is interesting because the special shape made for Lonsdale became a favorite and is still available today. (Same size).

# Chapter Seven

# BRITISH AND AMERICANS

During the second half of the nineties, a commercial war over the world's tobacco markets erupted between the two companies that commanded the largest tobacco businesses in England and the United States, namely the Imperial Tobacco Company, organized and directed by Henry Overton Wills, and the American Tobacco Company. At the end, both companies called it a draw and reached a settlement; each bought commercial rights over the brands and patents of the other. The American Tobacco Company (ATC) gained the right to appoint three executives and control of two thirds of the shares of a newly born consortium: the British American Tobacco Company would do busines in every possible market in the world except Great Britain, the United States and Puerto Rico, the latter two being considered territories of the ATC in the agreement.

This is the background that explains the links established between English and American companies based in Cuba around the turn of the Century. In time, these enterprises gained control over the majority of the country's cigar factories and a considerable number of plantations, stripping shops and warehouses.

British capital had arrived in Cuba since 1887, when Gustav Bock Muller, a German by birth who had settled in Havana in 1858, and his partner Benito Celorio, merged their factory called **El Aguila de Oro** and its 26 annexed brands, with Segundo Alvarez's **Henry Clay**, that had 18 annexed brands, to constitute the **Henry Clay & Bock and Company Ltd.** based in London. Gustav Bock, aside from being a shareholder of the company, remained as General Manager in Havana, while Segundo Alvarez, who reportedly got two and a half million pesos for his business, also remained as shareholder and Director-Manager of the **Henry Clay** factory.

Backed by British capital and with the support of competent personnel that he preserved, Bock expanded the business in 1895 when he bought two new cigar factories: **La Estrella**, with three annexed brands, **La Española**, with 16 brands, and leased a third factory, **La Intimidad**, with 3 annexed brands. In all, he was able to control 5 main brands and 66 annexed.

The English decided to invest in Cuban tobacco with the full official support of their government. They believed that British interests, being solely of a commercial nature, would be well received by both the Spaniards and Cubans who dealt with the growing and manufacture of tobacco. In Cuba, the **Henry Clay & Bock and Company Ltd.** was known to all as "the English company."

It has been pointed out that the tobacco industry was being affected by the serious economic crisis in the country as a result of Cuba's second war of independece (1895-98). The invasion of the West by General Antonio Maceo at the end of 1895 and first months of 1896 and his campaigns in the region had caused heavy damage in the tobacco plantations. In many localities of Pinar del Rio about 60 per cent of the growers had fled from the countryside mostly to join the rebel forces.

When the war ended in 1898, many factories faced insurmountable difficulties like shortages of raw material and lack of credits to continue in operation. Some factories were abandoned by their owners for political reasons, others decided to move to the United States and still others simply sold their businesses in view of the uncertain future that loomed ahead.

Under those circumstances, the English who were interested in expanding their investments in the industry founded the **Havana Cigar and Tobacco Factories Ltd.** in 1898, also called the **Havana Cigar Co.**, with Gustav Bock as their chief executive in the Cuban capital. This company bought four leading cigar brands and 35 annexed; these were **La Corona**, with 18, **La Rosa de Santiago** with 12, **La Flor de Naves** with 5 and **La Legitimidad** without any annexed brands.

On the other hand, American interests came into the Cuban cigar industry in 1899 during the military ocupation of the island at the end of the Spanish-Cuban-American War and the signing of The Paris Treaty in December, 1898.

The **Havana Commercial Company** of New York, founded with the purpose of buying cigar factories in Cuba and the United States, was financed by **H.B.Collins and Company**. Its first transaction in Cuba was to buy from **Francisco Garcia y Hermano**, their firm of tobacco warehouses in Havana and New York for half a million dollars. Garcia was appointed Managing Director in Havana and received the mission of buying factories and brands in the city. "Pancho" Garcia, as he was publicly known, was able to buy 12 leading brands and 149 annexed for the new enterprise: **La Antiguedad** with 6, **La Africana** with 10, **La Comercial** with 16, **La Meridiana** with 13, **La Rosa Aromatica** with 19, **La Flor de Inclan** with 2, **La Vencedora** with 11, **Manuel Garcia Alonso** with 36, A, de Villar y Villar with 2, and **El Siboney**, without any annexed brands. Not long after that, Garcia was replaced by Rafael Govin.

In 1902, the American Tobacco Co. (ATC), which was seemingly the main office of the Havana Commercial Co., sent one its highest ranking executives to conduct a series of buying transactions in Havana. It was then that Leopoldo Carvajal, Marquis of Pinar del Rio, and his numerous associates sold the venerable **Hija de Cabanas y Carvajal** with 8 annexed brands. This operation led to the founding of **H. de Cabanas y Carvajal Cigar Co.**, a subsidiary of the big American trust.

The same year, Jose Suarez Murias and his brother leased their factory with ten annexed brands to American investors who also depended on the American Tobacco Co. Consequently, by 1903 the ownership structure of almost the totality of the export quality cigar factories and their annexed brands showed the following pattern:

**Henry Clay and Bock and Co. Ltd**, with 71 brands, British.

**Havana Cigar and Tobacco Factories Ltd**, with 39 brands, British.

**Havana Commercial Co.**, with 161 brands, American.

**H. de Cabanas y Carvajal Cigar Co.**, with 9 brands, American.

**J.S. Murias Company**, with 11 brands, American.

Out of a total of 291 brands bought by foreign capitals, 110 were British and 181 American. As a whole, these foreign-owned firms came to be known as the **"Trust"** in the Cuban cigar industry.

The internal connections between these companies were not sufficiently clear then. A revealing clue became apparent a few years later, when the Trust's productive and commercial policies were jointly subordinated by Gustav Bock Muller, Director of the **Henry Clay and Bock and Co. Ltd.** where he was the top policy-maker until his death in 1910. In the end, Bock's firm finally got control of the majority of the shares of the other enterprises, which gave him full rights of management and decision. These companies bought the cigar, cigarrette and pipe tobacco factories with all their brands and facilities. They imposed the condition that all former proprietors and managers had to refrain from using their names in any identical or similar ventures they might undertake thereafter; nor could they appear as managers in any such ventures. In most cases this last restriction was also applicable to the partners who continued in the business. The British and American firms generally paid very large sums for the factories and brands they aquired, sums that the original owners perhaps never would have thought possible had the circumstances been different.

The new owners made it a point of keeping on the administrative personnel, overseers, buyers, salesmen, agents and occasionally even managers, for a number of years. Such was the case of Jose Rodriguez and Marcos Carvajal, from the **H. de Cabanas y Carvajal** administration. Rodriguez was exempted two years later and authorized to continue doing business in the tobacco industry.

He went on to be manager of **Partagas** first, and main partner at **Romeo y Julieta** later. Something similar happened with the Suarez Murias brothers, who had been kept on as salesmen in the new company and were exempted from the commitment twenty years later. Unwilling to let the opportunity pass him by, Jose Suarez Murias, at age 71, founded a new cigar factory: **La Radiante**.

Something totally different happened to those who breached the agreement and engaged in cigar ventures in which even slight reference to their names was made; the companies took immediate legal action against them. A "cause celebre" of this nature took place in 1932 when the **Havana Commercial Co.** filed a demand against the brands **Hijas de F. Villar Perez** and **Flor de Francisco Villar**, whose registration had been requested by Villar. The latter had been the owner of the **A. de Villar y Villar** factory which was bought by the **Havana Commercial Co.** in 1899. The company presented the contract, attested to by the Notary Public Francisco de Paula Rodriguez Acosta, which specified the obligation of the former owner not to use his name in any future tobacco enterprises that he might establish.

**In an effort to be different in looks, various colors, shades and designs were utilized by the major Havana brand leaders. (Same size).**

To administrate the plantations bought as of 1899, as well as all matters related to the agriculture and purchases of leaf tobacco, the Trust created the **Cuban Land and Leaf Tobacco company**. One of the golden rules of success in the manufacture of cigars states that the producer who controls his tobacco, from the plantation to the cigar-roller's bench, obtains stable quality results. Consequently, the newly acquired factories came over with numerous rustic farms and extensive plantations in the best tobacco lands of Cuba.

Following those well established practices, but inconsistent with the new principles of organization that were being introduced in the industry, the Trust also applied the system to agriculture thus creating a monopolistic structure in the tobacco industry.

A label from inside the box of GarrigayVilla cigars. (Same size)

The label from inside the box of Bock cigars. Both this label and the one above are reproduced in the same size as the original litho proof from which they were reproduced (into 4 colors in digital lithography).

# Chapter Eight
# THE INDEPENDENTS

In 1900 a number of cigar producers, almost all of Spanish origin, refused to sell their businesses and accepted the challenge posed by their powerful competitors. Their situation was extremely precarious, but they were willing to lift up the industry once again abiding by the same rules that had brought fame and prestige to their Havanas: competition based on quality and client satisfaction.

These men were neither afraid of the gigantic concentration of factories, brands, plantations, stripping shops and tobacco warehouses in the hands of the British and American trusts, nor of the enormous economic and financial resources these wielded. Their defiance was based on the fact that they knew every secret of an industry they themselves had built step by step. Furthermore, they mastered the particular characteristics of the foreign markets where their brands had jointly covered almost ten per cent of the total demand and won a lasting prestige.

The "independents," as this group of manufacturers was called, joined their efforts with the aim of keeping their rigurously hand-rolled cigars, made with the prime blends of Vuelta Abajo leaves, in the top rank as the best tobacco in world, a position they had conquered through skill and dedication.

In the forefront of the courageous enterprise were some of the best known brands: **H. Upmann**, still owned by one of the founding brothers; **Punch** and **La Vencedora**, of Manuel Lopez; **Hoyo de Monterrey, La Excepcion** and **Gener**, of the heirs of Jose Gener; **El Eden**, of Calixto Lopez; **Partagas**, of Ramon Cifuentes y Cia.; **La Miel**, of Tomas Diaz Valdes; **La Devesa**, of Pedro Murias; **Filoteo**, of Jose del Real; **La Sirena** and **La Venus**, of Manuel Rodriguez; **La Convencion**, of Jose Diaz Rodriguez; **El Mapa Mundi**, of J.F. Berndes; **La Diligencia** of Bernardo Moreda; **C.E.Beck**, of C.E.Beck y Cia.; **Por Larranaga**, of Eustaquio Alonso y Cia.; **Romeo y Julieta**, of Jose Rodriguez y Cia.; **Fonseca**, of F.E.Fonseca; **Sol** and **Luis Marx**, of F. Beherens; **La Flor de P.A. Estanillo**, of Pedro A. Estanillo; **El Crepusculo**, of Jose Rocha and Rafael Garcia Marques; **La Cruz Roja**, of Rabel, Costa y Cia.; **El Rico Habano**, of B. Menendez y

**A.R.Suarez made a more modern label featuring a beauty of the 1920's. (Same size).**

**The label from a small box of Santa Felipa cigars from Ramon Fernandez. (Same size).**

A very interesting label from Manuel Alvarez's cigar box.

Hermano; **La Flor de Fernandez Garcia** and **La Belinda**, of Francisco Menéndez Martinez.

In the following years, the tenacity shown by the "independents" in pursuit of a revival of their businesses led them to auspicious results. Their brands stood side by side with the "Trust's" products and recaptured the smokers' preference in numerous foreign markets. Official statistic reports of the Ministry of Finance show that already in 1904, the independent factories supplied 48 per cent of all cigar exports that year.

The label from inside a box of Flor de Alvarez, featuring Sr. Alvarez himself. (Same size).

This six color lithographed label, gold stamped and embossed is printed to look like Cuban cedar. Actually it is just printed on common coated one side litho paper (100 gsm which means it weighs 100 grams per square meter which is about 70 pounds for 500 sheets of 25 x 38" ). (Same size).

One of the many fakes sold as Havana cigars was this La Carolina. The Jose Alonso y Ca. bought the tobacco from Cuba but the cigars were rolled (and blended as they were rolled) in the U.S. (Enlarged 50%).

# Chapter Nine
# INDEPENDENTS VS. TRUSTS

In 1904 the rearrangement of the industry during its initial stage defined two different camps. On one side, the Trust, with the factories and brands that had yielded to British and American investors; on the other, the manufacturers who chose to remain on the island and stay in business with their own factories and registered brands. This last group, which has already been described, remained competitive on the market where its brands were highly respected, and some of its members even managed to gain a solid position and extend their reach.

Although an outright trade war did not actually occur, competition inevitably breeds rivalries and clashes, which in this instance were frequent and repetitous.

In July, 1904, the trust's Board of Directors published a booklet in New York under the title *The Truth about Havana Cigars*, written by Gustav Bock, whose origins and history

**An exceptionally well printed label from a box of Perez y Diaz made by the successors of Manuel Lopez. (Same size).**

**This strange label with text in English, Spanish and French would lead you to believe that the man's name was *F.P. del Rio*. (Same size).**

within the tobacco industry we have already traced. The booklet, widely distributed in the United States, Great Britain, France, Spain, Germany, Austria and other European and American countries with a market for Havana cigars, caused an irate reaction by the independent manufacturers.

The document appears to have accomplished its intended purpose of proclaiming far and wide the supremacy of the cigar brands manufactured by the Trust.

In the booklet, Bock adopted a supposedly "personal" stance based on his 46 years of experience in the trade. One of the paragraphs read as follows:

*My statements are catergorical and can be confirmed, and I therefore have the right to demand they be accepted as true.*

Among other things, he claimed to have discovered the best method, down to the smallest detail, for manufacturing the finest high quality Havana cigars. Based on this assumption, **"Don Gustavo"** expounded his arguments with the certainty of a businessman totally convinced of the merits of his product:

*If what I have just said is understood and accepted, no further proof is needed to demonstrate that the cigars manufactured under my personal guidance at the 23 factories owned by the* **Havana Tobacco Company**

59

**The Cristobal Diaz box label features beautiful birds with a nest containing eggs along with many flowers. (Enlarged 50%).**

**A label used for various local cigars from Pinar del Rio. (Same size).**

Finca AJICONAL. - Pinar del Rio.

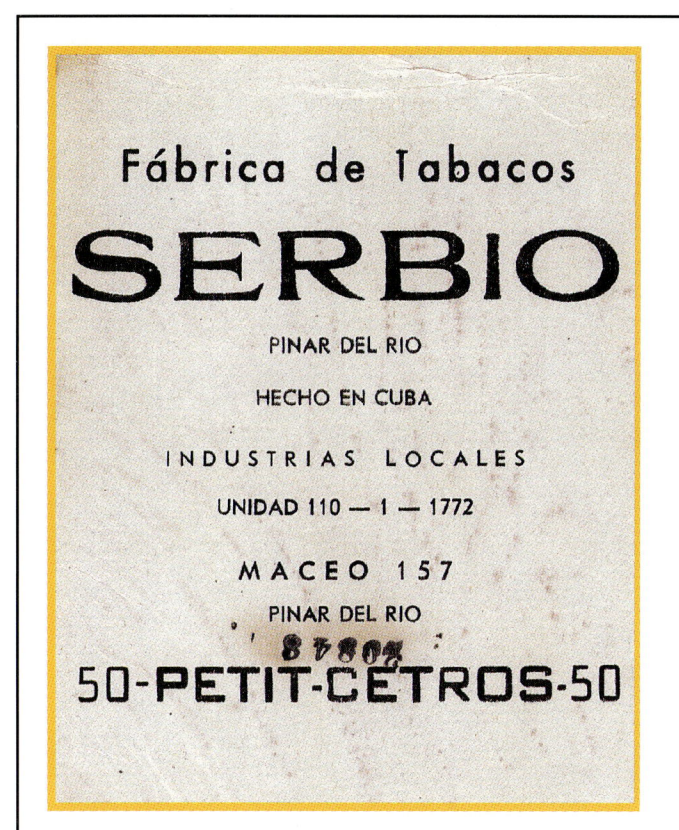

**A plain poorly printed wrapper used for machine-made cigars produced in the city of Pinar del Rio.**

*can by no means be equalled in aroma, quality or craftsmanship by any other cigar in the world.*

Apparently, independent manufacturers simply ignored these remarks. But no doubt privately stung, they sought their own advocate. In August 1905 another booklet was released in Spanish and English offering –like Bock's– the "personal" views of its author, the controversial Jose Gonzalez Aguirre, an experienced cigar roller, journalist and founder of the first magazine on tobacco affairs published in Havana.

After pointing out how the trust sang its own praises, Aguirre fell implacably upon Don Gustavo. He claimed that the lowliest craftsman in Havana could easily counter the arguments presented by the old businessman. According to Aguirre, no one in Havana thought Mr. Bock intelligent enough to manage a cigar factory. To reinforce his dissection of Bock, Aguirre cited both old and new manufacturers, warehouse keepers, merchants and contractors in tobacco exports who considered Bock *an active, skillful and smart salesman who knows European and American markets well, and the kind of cigar and* **vitola** *to offer each. Regarding agriculture and industry–read the booklet–he lies well short of average.*

Acknowledging Bock's achievements in business, Aguirre attributed them to the fact that he always recruited expert advisors such as Manuel Valle, Melchor Fernandez, Benito Celorio and others. As proof of this remark, Aguirre reminded the readers that when Bock had lacked such support and was left on his own, he had been forced to lease his factory **El Aguila de Oro** and concentrate on contracting, an occupation at which he excelled. Aguirre stressed that the practice of leaving experienced people in key positions was one of Bock's most successful abilities, with proven merits during the establishment of the British "union" that took over **Henry Clay**, **Aguila de Oro**, **La Corona**, **La Espanola**, **La Estrella**, **La Rosa de Santiago**, **Don Quijote**, and **La Intimidad**, when some of the previous owners remained in managerial positions. Likewise, he gave Bock credit for his initiative during the critical days of the war from 1895 to 1899.

Yet Aguirre's rebuttal included elements that undermined Bock's proclaimed authority. He stated that when British and American interests merged to form the "trust," Bock's authority had vanished as a result of new management practices and a blatantly damaging reorganization of the industry. The Board of Directors was based in New York, and, according to Aguirre, intended to oversee the whole business–from production and manufacturing down to consumption–from that vantage point. Since Bock opposed the change, he was stripped of authority, while other bosses were sent out to fill more important decision-making positions. Although he formally appeared as Director General of offices, factories, warehouses and plantations, Bock neither managed nor decided

**The factory of dynamic cigars, *Fabrica de Tabacos el Dinamico*, located on Animas 59, Havana, produced this vanity label.**

The beautifully lithographed label which appeared on the sides of the square boxes of these Francisco Perez del Rio cigars. (Same size).
Below is another inside-the-box label of F.P. del Rio cigars which has the same basic design. (Enlarged 15%).

A beautiful six color label, gold embossed, for Santa Damiana cigars. (Enlarged 95%).

anything because the real boss giving the orders was Mr. Seidenberg in New York.

In this context Aguirre analyzed some of the organizational changes he considered unwise. By then, the 23 factories Bock boasted had shrunk to 11, while at the time the rebuttal was released in 1905 the number had dropped even further to 7 as the policy of concentration and centralization gained ground. These were: **Cabanas y Carvajal**, **Henry Clay**, **La Intimidad**, **El Aguila de Oro**, **Suarez Murias**, **Manuel Garcia**, and **Carolina**. The rest had been merged and turned into annexed brands.

The location of a cigar factory has always been considered an important environmental factor affecting the quality of the product. It is not only a matter of comfort in the facilities that favorably disposes the rollers for their creative work, but has to do with the peculiar atmosphere in which the many delicate tasks preceding and following the **torcido** can achieve peak results. This notion was so deeply rooted that when the Trust moved its factory to New Jersey, it tried to replicate the facilities at the former *Palacio de Aldama* in Havana, where **La Corona** was based, down to the smallest detail, from tools to lighting and temperature.

Thus, in his criticism of what he viewed as ill-advised steps taken by the Trust, Aguirre declared that out of the 7 Havana factories mentioned, the only ones still occupying their original premises were **J.S. Murias**, **La Intimidad** (which could not be moved because it was leased), and **Henry Clay** with its annexed brands **La Corona**, **Rosa de Santiago**, **La Espanola**, **Don Quijote**, **Estrella** and **Manuel Alonso**, along with **La Vencedora**, **La Antiguedad** and **La Rosa Aromatica**.

Aguirre noted that the original spacious quarters occupied by **El Aguila de Oro** were vacated and operations moved to a new location at No.2 Belascoain St., in order to concentrate general stripping and drying in a single place from where the tobacco leaves were later distributed to the various factories.

*At the Trust's factories,* he wrote, *materials, preparation and blends are one and the same for all factories, at least those located in Havana, since none of them have their own warehouse, or stripping and drying operations.*

*...In addition to general stripping, drying is also centrally performed, from where large crates of exactly the same fillers are distrib-*

La Diana labels. Manufacturers of cigars often used their children or grandchildren as models for their labels. (Same size).

Established in 1848, *El Rey del Mundo*, or King of the World, cigars were a huge success. (Enlarged 50%).

*uted daily to the various factories,* remarked Aguirre.

The difference between brands was therefore pure imagination, and their only truly distinctive features were the seals on the **vitolas** and the prices charged to the consumers.

The critic, however, acknowledged the fact that by owning fine tobacco plantations in Vuelta Abajo, Vuelta Arriba and Partido, the Trust was in a position to offer guaranteed purity in the origin of its leaves.

Moreover, to substantiate his argument in favor of rigorous time-honored procedures, Aguirre cited experienced captains of the industry and their most recent followers such as Gonzalez del Valle, Julian Alvarez, Perez del Rio, Jaime Partagas, Manuel Lopez, Pedro Bances, Antonio Rivero, Jose Gener, Manuel Valle, Segundo Alvarez, Cifuentes y Fernandez, Calixto Lopez, Villamil y Rivero, and many others adding up to a total of 43.

They were all founders and champions of manufacturing traditions that represented the finest quality among Havana cigar brands with names such as **H. Upmann**, **Eden**, **La Excepcion**, **Partagas y Cia.**, **Ramon Allones**, **Romeo y Julieta**, **Flor de Tomas Gutierrez**, **Por Larranaga**, **Punch**, **High Life**, **Flor el Todo**, **Lord Beaconsfield**, **El Rey del Mundo**, **Castaneda**, **Modelo de Cuba**, **Redencion**, **Sol**, **El Guardian**, **Belinda**, **La Devesa**, **Newton**, **Flor de C. E. Beck**, **La Sirena**, **La**

Vanity rings by Romeo y Julieta. (Enlarged 10%).

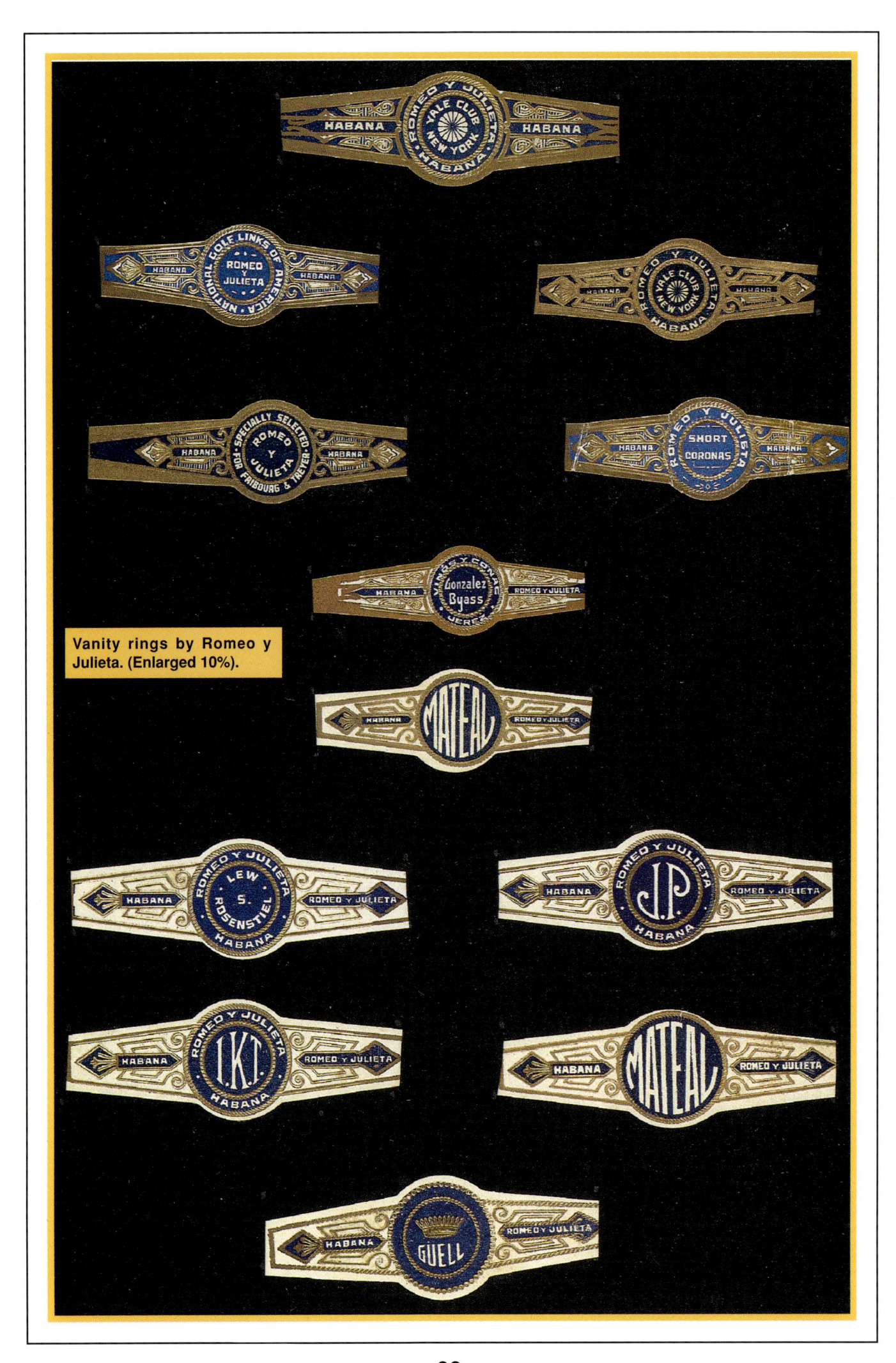

Vanity rings from El Rey del Mundo including some blanks awaiting imprinting. (Same size).

**Capitana**, **Filoteo**, **La Diligencia**, **El Crepusculo**, **El Puro Habano**, **La Miel**, **El Ecuador**, **La Flor Cubana**, and **La Angelica**.

Another point made by Aguirre was that the Trust's claim to supreme quality was totally unwarranted. He countered that the distinctive traits of each brand lie in the procedures followed by the various factories in arranging their materials according to the maturity of the leaf, the precise timing for the mixtures, the sensitivity and care afforded the blends to obtain a particular taste and aroma, and the touch and skill of each manufacturer. He maintained that variety was preserved by the independent factories where traditional procedures were strictly followed, as opposed to the Trust's factories where preparation and blends were one and the same for all.

Finally, he convincingly argued against Bock's assertion that 80% of Havana cigars sold throughout the world carried brands manufactured in the factories under his stewardship. Based on figures obtained from the Havana Office of Customs Exports, the only point of exit for Havana cigars, Aguirre showed how in 1904 Trust brands only represented 53% of total exports, which meant there was a certain balance between the Trust and the independent manufacturers, both in volume and in value.

This controversy, which illustrates the existing rivalries, generated national sympathy for the manufacturers who had rebuilt a large part of the Cuban tobacco industry and were considered the custodians of the purest manufacturing traditions.

Rings made in various shapes.

# Chapter Ten
# MECHANIZATION OF THE INDUSTRY

The first attempts at mechanization in the Cuban tobacco industry were made in 1925. Ever since the crisis and recovery of the late 19th and early 20th Centuries, certain tools and machines were introduced into some stages of the process, but they were by nature compatible with the crafts traditionally used in manufacturing, contributing to the ease, speed and precision of the craftsman's task without ever intending to replace the artisan. The new machines were generally used in loading, storage, aging and fermentation, and were more readily accepted in the production of cigarettes rather than cigars.

In 1925, the **Compania Tabacalera Nacional Habana, S.A.** was founded in Havana under the joint sponsorship of **Jose Diaz Villamil, Benito Santalla and Enrique Berenguer**, who were also shareholders of Por Larranaga and had obtained a contract with the North American Machine & Foundry Co. to operate cigar manufacturing machines in Cuba in the same way they were being used in the United States. The agreement included a royalty payable to the American company for each thousand cigars produced. **Tabacalera Nacional** manufactured the cigars and passed them on to the well-known **Por Larranaga** for marketing.

Initial reaction was one of natural curiosity and caution in the presence of something new. Progress in the industry was followed with equal admiration and distrust. The fact that the workers did not show hostility toward a machine taking over operations which had previously been the exclusive domain of the rollers might have stemmed from their conviction that the experiment would fail. The sponsoring company even set up exhibitions in an attempt to promote the new procedure. A particularly striking novelty was the introduction of women workers as machine operators.

As mechanized output stabilized, signs of rejection and unrest grew steadily. The **Por Larranaga** workers went on strike and the unions boycotted their products, while work-

The famous Por Larrañaga box label.

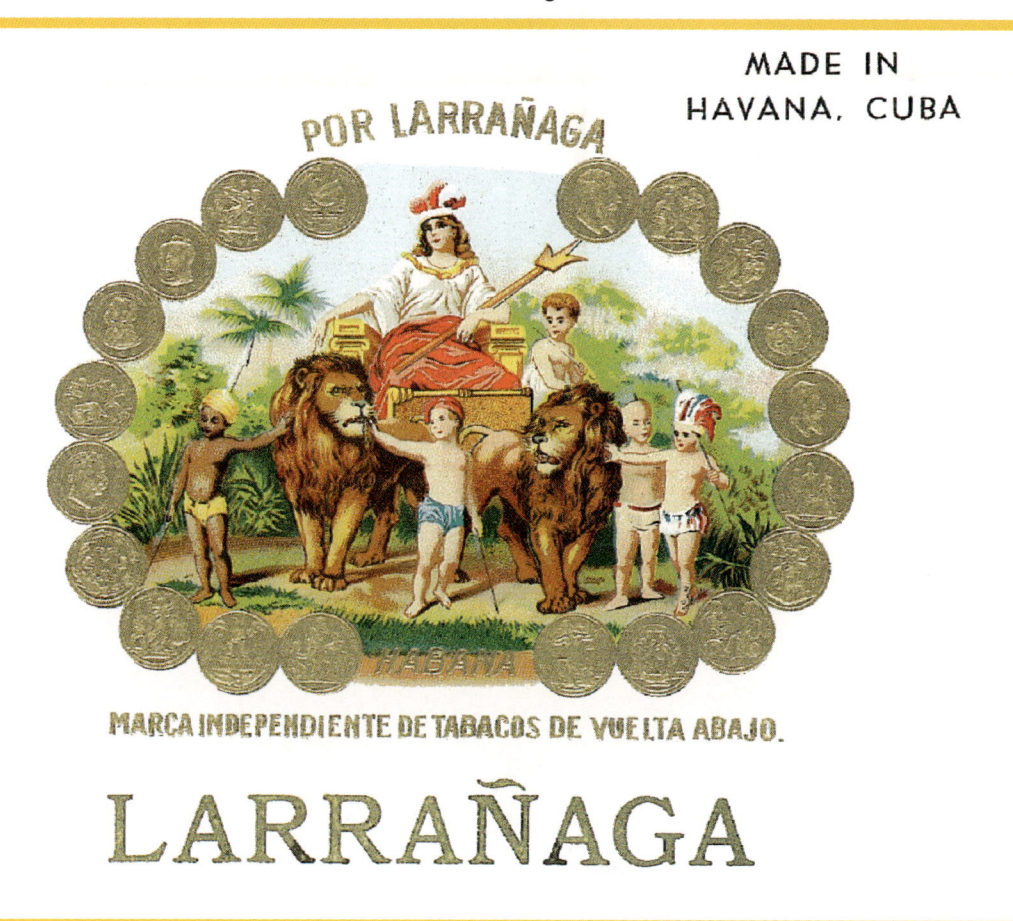

The brand Elios was manufactured by Lopez, Fernandez y Cia in Cabaiguan, Cuba even though the label proclaims it is *puro Habana*! (Same size).

Another *Elios* label but this one doesn't claim to be pure Havana. (Same size).

This lovely label was issued by Alvarez, Lopez y Ca. from Havana. (Same size).

ers from other sectors, organizations and most of public opinion joined the powerful campaign.

Even manufacturers repudiated the change, although their view points were quite different from that of the workers. In October 1926, **La Union de Fabricantes de Tabacos y Cigarros de la Isla de Cuba** addressed the President of the Republic denouncing the problems created by mechanization and calling for the establishment of a top level committee to assess and determine the situation, taking into account internal and external factors that could by no means be ignored.

Underlying the arguments against the introduction of machines presented by independent manufacturers was their lack of funds to undertake such a vastly complex change, which set them at a serious disadvantage with the Trust. In 1926, Francisco Pego Pita, president of the **Union de Fabricantes**, issued a detailed report on the effects of mechanization on workers, manufacturers and the national economy. He blamed the drop in exports on imitation Havana cigars manufactured abroad, lack of adequate promotion, high prices for hand made cigars and the resulting decline in sales. In contrast, he pointed out that the **puro habano** was still the one in highest demand on the market.

In specific reference to the introduction of machines, Mr. Pita declared that it would make no difference whatsoever given the distinctive high quality of hand made cigars, which depended mainly on the irreplaceable skills of the rollers. Therefore, he concluded, what was needed was a policy aimed at bolstering the Havana cigar to include, among other things, constant vigilance against imitations and renewed efforts to lower tariffs around the world.

Mechanization was not only damaging to the workers it would eventually lay off, but to the manufacturers who could simply not afford it. The industry would increasingly shrink, causing massive unemployment. Domestic consumers would not benefit either, since machine made cigars could never be less expensive than the popular "one cent" variety hand crafted by unemployed rollers in times of crisis.

Among the irrefutable facts quoted by manufacturers were figures showing how it would take 300 rolling machines at a cost of 2,500,000 pesos to produce 70% of the cigars made by hand at the time. Royalties owed to machine suppliers would amount to 480,000 pesos a year, all of which would benefit foreign interests. In addition, there would be unavoidable expenses on spare parts and replacements as well as maintenance and specialized personnel. Furthermore, they pointed out that the making of machines was in itself an industry generating jobs abroad while making thousands of workers redundant at home.

As a result of the struggle against the introduction of machines in the making of cigars, on 28 February 1927 a Presidential Decree was issued requiring all machine made cigars to carry a distinctive stamp, and every box to be sealed with a special tape specifying the method used in their manufacture, thus giving consumers a choice.

Viewing the measure as moderate and half-hearted, the **Federacion Nacional de Torcedores de Cuba** urged the **Comision Nacional de Propaganda y Defensa del Tabaco Habano** to lobby the executive branch for more drastic action, arguing that machines would lead the Cuban tobacco industry to its ruin.

**Por Larranaga** continued to fend off the onslaught. In 1928 it addressed the Secretary of Agriculture, Commerce and Labor, explaining that machine made cigars were well crafted and excellently packaged, while the expensive **vitolas** (referred to as **regalias**) would always require the irreplaceable skills of the roller. It maintained that the introduction of machines, far from damaging cigar makers, worked to their advantage.

The issue was widely debated, since not only the **Comision Nacional de Propaganda y Defensa**, but also warehouse owners, growers, factory workers, congressional committees, civic organizations, municipal governments and entities with no vested interest, became involved.

In September of 1928, Jose Mujica as president and Jose Aguirre Lopez as secretary of the **Federacion Nacional de Torcedores de Cuba** explicitly called on the President of the Republic to ban the use of cigar-making machines in Cuba. In July, 1929, **Por Larranaga** demanded that Decree 266 of February, 1927 enforcing the distinctive stamps and seals be repealed. The heated debate even reached Congress where several bills contemplating prohibition were circulated.

**Por Larranaga** undertook to present the **Comision de Propaganda y Defensa** with monthly reports on the trade performance of

Calixto Lopez was a very important cigar maker in Havana. Their label featured their factory. (Enlarged 15%).

This La Preferencia label is printed on paper to look like wood. (Same size).

machine made cigars. When in November, 1936 the report failed to arrive, the committee demanded the document. Several months later, in June, 1937, **Por Larranaga** explained in its reply that since the previous report in 1936, the market for machine made cigars remained paralyzed. It also notified the committee that an agreement had been reached with **American Machine & Foundry** to withdraw the controversial machines from Cuba.

On July 15th, 1937, the representative of the **Federacion Tabacalera** on the **Comision de Propaganda y Defensa del Habano** notified that body that the **Compania Tabacalera Nacional Habana S.A.** had returned the machines to the owner in the United States. The issue was thus closed and the first attempts at mechanizing the Cuban cigar industry ended.

Arguments for and against the use of machinery, however, had a common denominator. The drop in export levels for rolled cigars, and at times even leaf exports, due to increased tariffs, substantially raised the retail price of the product. The notion of introducing new methods to cut production costs by means of cheaper cigars was aimed at reducing consumer prices and drawing low income smokers into the market, thus increasing exports.

When in 1927 the **Comision de Propaganda y Defensa del Tabaco Habano** was

Dr. Jose Manuel Cortina.

established, Dr. Jose M. Cortina, who had introduced the bill in Congress, had maintained that given the worldwide reputation of Cuban cigars as a luxury product and privilege of the wealthy, low income markets could not be pursued with a cheaper product because in the stiff competition that would follow, the Cuban industry would undoubtedly operate at a disadvantage. He argued that Cuba should make use of the excellence of Havana cigars to increase purchasing power, thus extending its use beyond the very wealthy.

Mechanization, naturally viewed with sympathy among foreign interests linked to the Cuban tobacco industry, was intended to increase productivity. Profit from the excellent reputation of Havana cigars through low cost manufacturing (in their opinion) would not endanger traditional cigar makers because it targeted a new group of consumers.

Nevertheless, the issue was not dead and forgotten within the tobacco industry. Exports remained at a serious low with no immediate improvement in sight. Hopes were set on the **Comision de Propaganda y Defensa**, which almost at birth faced the world economic crisis

High class vanity rings from Partagas.

**Lord Byron is proclaimed in this vanity label. (Enlarged 20%).**

of 1929 and the resulting general collapse. By 1932 cigar exports, including those destined for the US market, had dropped 60%.

Clashes between growers and manufacturers were marked by arguments in which the growers defended their right to export leaves while the manufacturers advocated industrial applications and development. World War II brought about a significant reactivation of markets for Cuba and export manufacturers did not want to miss out on the bonanza. Faced with the reality of expanding markets for machine made cigars throughout the world, they became increasingly convinced that change was needed to sweep aside time-honored practices. In 1944 export manufacturers spoke out in favor of reintroducing machines in cigar manufacturing. In 1945 a special committee examined the issue, but animosity soared once again with renewed force. Final decision was deferred until 1950.

A small sector within the industry adopted mechanization, after workers grudgingly yielded to the overpowering reality of the markets. In 1953, 50 million machine made cigars were produced in Cuba, slightly over 13% of a total 375 million.

Mechanization finally prevailed as the result of a gradual and painful transformation by means of which the industrial process was split along conceptual and practical lines. Low cost inexpensive machine made cigars were manufactured for the domestic market, while all those intended for export were still traditionally hand made in order to preserve the reputation earned by Havana cigars abroad on the assumption that the key to quality lies in the artistry of the master craftsman.

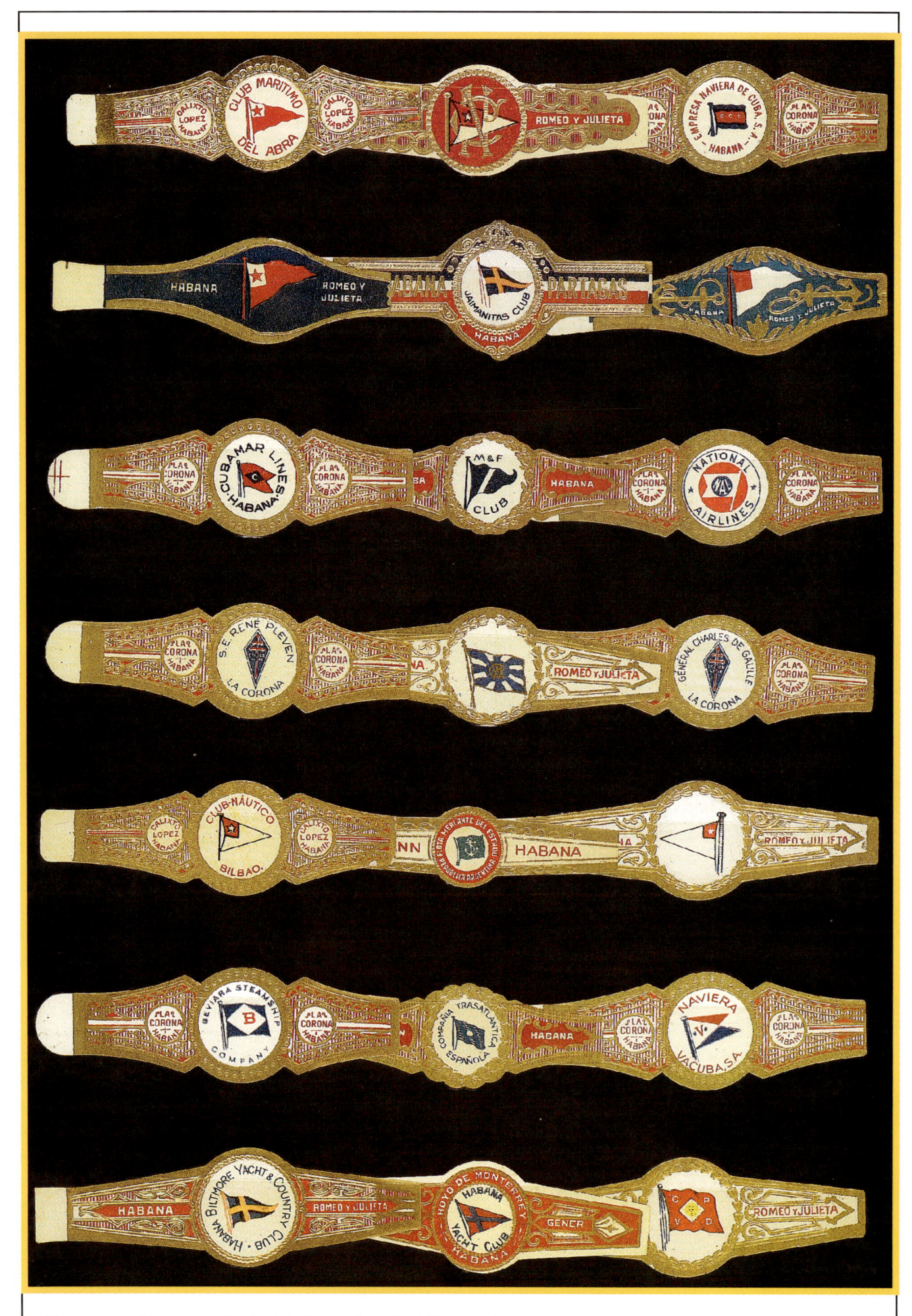

**Many steamship companies had their own *brands* of Havana cigars made with their vanity bands. (Enlarged 10%).**

# Chapter Eleven
# THE INDUSTRY SINCE THE THIRTIES

During the thirties and early forties of this Century, the cigar industry was faced with another crisis situation. The sharp decline in the international markets forced the manufacturers to give serious consideration to the terrain the competition was gaining in those markets. Internally, the debate on the need to cut costs in the industry to make it more competitive was mounting. Thus the idea of cutting wages and of installing machines was not completely ruled out in cigar manufacturing circles. The withdrawal of the foreign firms in 1932, when they took their brands to the United States to manufacture cigars there, made them possible invaders of the markets which the general economic crisis was keeping the Cuba-based industry from reanimating.

The development of the Trust in the United States faced the local industry with the need to cut costs which, they argued, were influenced by the high wages earned by the workers, the high customs duties Cuban cigars had to pay to enter the United States, plus the expense implicit in the "smokes" (the cigar allowance to which the workers were entitled). The customs duties at the time were estimated to be eleven times greater on cigars than on the tobacco that went into them. When the Trust found it impossible to push through the 12% wage cut its representatives were seeking and subsequently ran up against the firm refusal of the workers to have their smokes cut, the most effective expedient it could come up with was to transfer their brands to the United States and import the tobacco from Cuba to manufacture their export cigars under less costly and more profitable conditions. This was indispensable if they were to have access to a greater number of

**Bands, or rings, made for various vanity purposes.**

consumers because amid the conditions of the economic crisis facing the United States and other countries, whose markets had contracted very significantly, the high retail price of Havana cigars made it practically impossible to increase sales. This was the assessment made by the Trust that transferred the "independent" manufacturers. They were *ready to cooperate to test the strengthen of the Federation of Cigar Makers.*

After a strike that lasted six months, the manufacturers succeeded in opening subsidiaries in towns near Havana, such as Bejucal, Calabazar and Santiago de las Vegas, where a large number of unemployed cigar makers agreed to work for the reduced wages.

Finally the workers arrived at an agreement with their bosses. However, George Washington Hill, the top executive of the Trust, had been working to manufacture the Trust's export cigars in Trenton, New Jersey. To do so, he had ordered the construction of a building with all the requirements and conditions of the factories in Cuba and proceeded to train some 2,000 local women.

He also took with him one Jose Garcia, who had for many years been the blend master for La Corona. When the facility and other conditions were ready, he began to manufacture cigars under the La Corona and other brands, but always with an indication of the origin.

Thus Trenton, New Jersey, began to produce a Cuban product at a site in which everything was copied from the original: the heat

**Bands with the names of the vitolas (sizes, shapes) of various manufacturers.**

Vanity rings which are, in effect, endorsements. (Same size).

High class rings which were for prestigious customers. (Enlarged 15%).

level, the lighting conditions existing in Havana, the work shops and the general environment of the premises where La Corona and other brands were manufactured. The start-up of manufacturing operations in the territory of the United States freed the manufacturers of the heavy import duties and gave them certain other advantages in the retail trade, all of which allowed them to reduce the end price of the cigars. The procedure used in the work shops was the classical hand rolling method. This event was a turning point that condemned the island to the role of producer of raw materials for the manufacturing operations established in the United States.

The advertising campaign unleashed in the new conditions that developed in the cigar industry revealed the true entrails of this great maneuver whose objectives became quite clear in the ads that appeared in the press: **Now you can buy three Coronas for a dollar, and they are better made than cigars rolled in Havana.**

Bachelor Maradona brands on cigars manufactured in subsidiaries outside of Cuba.

The Henry Clay & Bock Trust had the initiative of setting up a Cuban company to use its registered brands. They suspended production for export in Cuba and gave Tabacalera Cubana S.A., which was located at No. 106 Agramonte Street, in Havana, all their operations for the domestic market. Producing cigars for export in Cuba was not profitable. In the opinion of the Trust, the only alternative was to get out of the cigar business or transfer operations to the United States using imported tobacco grown in Vuelta Abajo.

Many voices were rallying with renewed force to demand special attention for the country's second largest industry. Huge volumes of leaf tobacco were crowding the warehouses and creating an over-production glut. The lack of sales for both leaf tobacco and cigars was raising the specter of a collapse of enormous proportions.

Representatives of the province of Pinar del

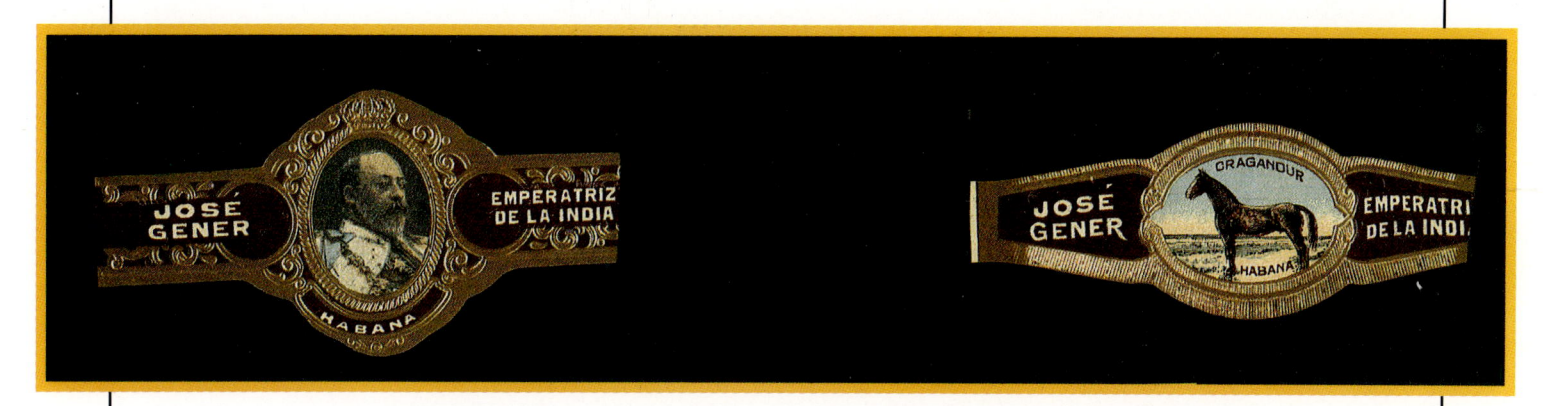

Kings, winners of elections and horse races, often adorned vanity rings. (Same size).

The Trust was thus able to introduce a less expensive cigar into its most important market while the genuine Havana cigar was being strangled by tariff barriers and by the power of the foreign monopolies.

The indication of the origin of the cigars made in the United States stemmed from a decision of the Cuban State which had, ever since 1928, forbidden a number of brands, including la **Flor de Cuba**, **La Rosa Aromatica** and **La Flor de Murias**, belonging to Havana Commercial Co., to use the words Havana-Cuba in their packaging as of the moment they transferred their manufacturing facilities to Tampa, Florida. The Cuban State had also ordered (and the proprietors complied) the cancellation of the license to use the **La Devesa** brand, by Eduardo Suarez Murias, as well as the El Mercurio, Diaz y Garcia, La Popular and

Rio introduced a number of bills in the legislature demanding a limitation of the tobacco growing areas as well as the corresponding restrictions, but they were thwarted by the organization of growers and the import companies that were benefiting from the low price of leaf tobacco.

The controversy became particularly heated when the opposition, the tobacco traders and importers, started to turn their arguments in the direction of the Government of the country. Their lukeqarm or non-existent action in favor of the signing of new trade treaties or to protect and promote export markets, received from the public the reaction it deserved. When the renewed Trade Reciprocity Agreement was signed between Cuba and the United States in 1934, cigars were once again absent from all its benefit clauses.

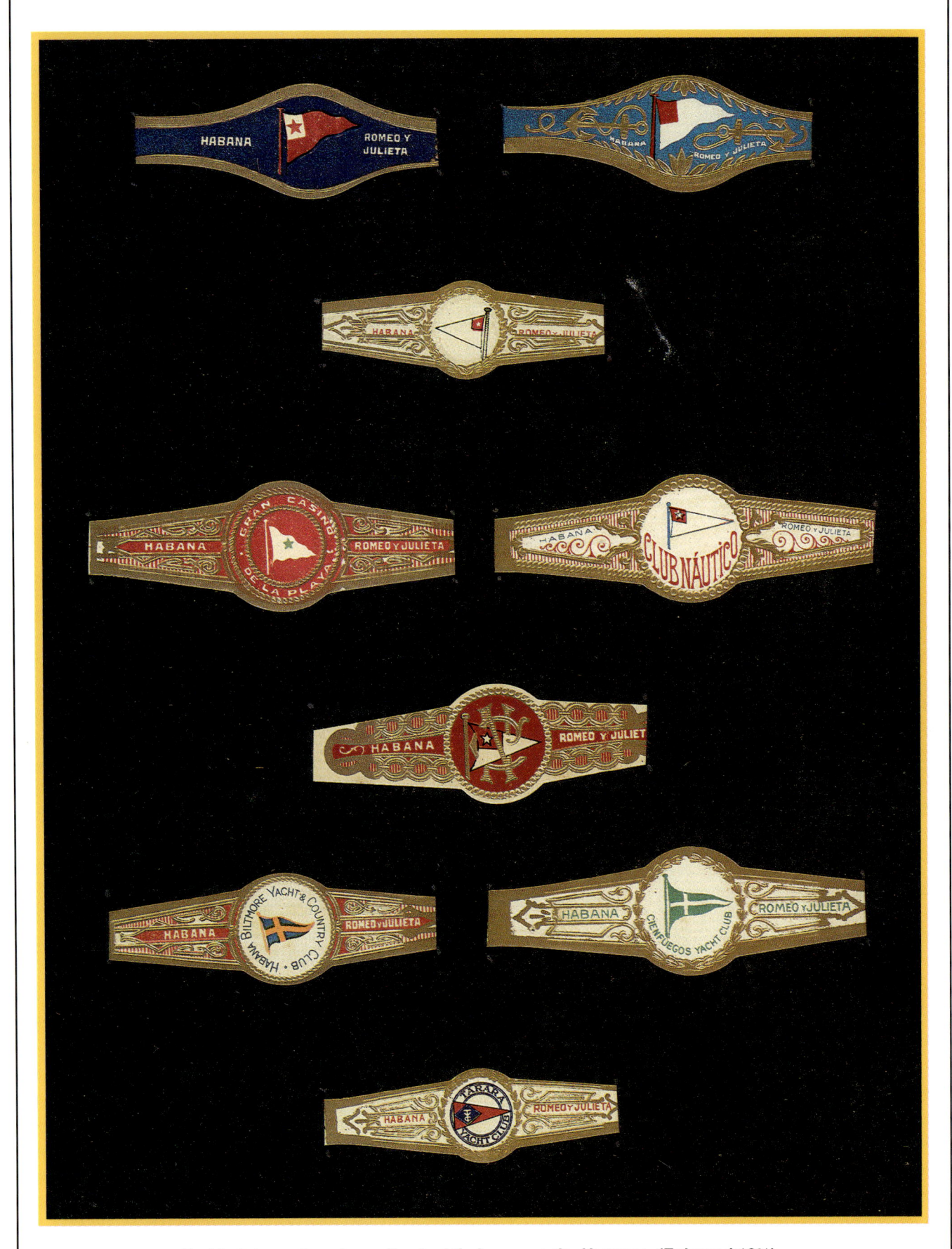

Yacht clubs and marinas often had their own vanity Havanas. (Enlarged 10%).

The decade of the forties was ushered in by the Cuban cigar manufacturers with a spirit of defeat. But the forties also brought with them the Second World War and as happened in the previous war, the initial reaction was a profound depression. In 1940, Cuba had the lowest cigar export performance ever recorded since the consolidation of the industry in the 19th Century: 14,000,000 cigars. In 1941, exports increased to 20,000,000 units and in 1942 the figure was 17,800,000, the second lowest figure in the Century since the recovery of the industry in the 1902-1906 period when exports hovered around 200,000,000 units. A record 257,776,239 cigars were exported in 1906. In the following years and up to 1913, exports ran from 189,000,000 in 1908 to 184,000,000 in 1913. In 1914, the year the First World War broke out, exports dropped to 132,000,000 units, 52,000,000 less than the previous year. The decline continued to 121,000,000 in 1915, 122,727,000 in 1916 and 112,000,000 in 1917. With the end of the war came a recovery. Exports in the period between 1918 and 1920 totalled 149,000,000, 162,000,000 and 150,000,000 cigars, respectively. The average for the next seven years was in the vicinity of 83,000,000, except for 1925, when 108,000,000 cigars were exported. The downward trend continued between 1928 and 1933, with an average of 63,363,000, and in the period leading up to the Second World War, 1934 to 1938, exports dropped even lower, with a yearly average of 38,000,000, up to 1939, when the figure was 29,000,000.

Very specific markets, such as Great Britain, which in the pre-war period (1935-1939) was taking some 20,000,000 cigars a year, closed

**Often vanity rings had crests of arms and other sophisticated designs which makers often produced for their own account. (Same size).**

down completely, a situation that remained unchanged for several years after the end of the war, when the United Kingdom increased its consumption of cigars made in its colonies, particularly Jamaica, alleging that it lacked the hard currency needed to acquire them in Cuba. This was a hard blow for the Cuban industry, as sales to England at the time accounted for 58.72% of all Cuban cigar exports, plus the fact that in January, 1937, England had signed a Trade Agreement with Cuba whereby the many British interests established in Cuba, mainly insurance and railroad companies, benefitted from concessions and advantages given in exchange for the purchase of Cuban cigars. The remaining European markets were also blocked off by the armed conflict.

The situation took an unexpected turn for the best when the United States market picked up between 1942 and 1944. Imports of Cuban cigars increased from around 18,000,000 to 181,000,000. That figure dropped back to 65,000,000 in 1943 and rebounded to 110,000,000 in 1945. Of total cigar exports during those years, the United States accounted for 23,000,000 in 1943, 47,000,000 in 1944 and almost 62,000,000 in 1945.

The fact that the recovery was transitory became evident in 1946, when Cuban cigar exports virtually collapsed to around 60,000,000 units. The ensuing chaos did away with all internal discrepancies. The United States recovered the rates and levels of its industrial production and again forced the retreat of Cuban cigars while it favored imports of leaf tobacco. This produced a confrontation between tobacco growers and cigar manufacturers in Cuba: the latter maintained that the

**Vanity rings showing a bank in Hoboken and the crown of India. (Same size).**

84

Vanity rings of Calixto Lopez. (Same size).

country had to defend the future of its industry; but not the future of a ruined, backward and miserable industry, argued the former.

Nevertheless, the restriction on growers finally came by way of a curious arrangement based on the creation of the Stabilization Fund (financed by a price increase on cigarettes), which ensured a lower domestic price for leaf tobacco, while fostering higher prices for the export market. Enforcement of the agreements was left in the hands of the National Commission for the Advertising and Defense of Cuban Cigars and of the Stabilization Fund, whose coffers were misappropriated. The result was a series of scandals involving corruption and favoritism and the most onerous consequences fell on the shoulders of the hard-working growers.

The Commission for the Advertising and Defense of Cuban Cigars was created by virtue of Law 12 of June, 1927, on an initiative promoted by Dr. Jose M. Cortina, legislator and judge, who for many years was the legal representative of the tobacco growers.

It was organized with representatives of the production and industrial sectors and was presided over by the Secretary, or Minister, of Agriculture, who represented the Government.

The Commission had five committees with the following areas of responsibility: advertising, growing, defense, budgets and auctions.

During the years it operated, and prior to the crucial times described above, it had helped in work aimed at improving the systems used in planting, culture, harvesting, curing and packaging, as well as in the introduction of new varieties, pest control, soil analysis, meteorological services, publication of agricultural bulletins, distribution of seeds especially selected for particular soils, reports on fertilizers, harvest percentages and tobacco diseases. It opened its Experimental Station in San Juan y Martinez, Vuelta Abajo, with its own fields, laboratory, library, weather facility and radio station. The Experimental Station had subsidiaries in Las Villas and in Oriente provinces.

In its work abroad, it maintained a permanent institutional advertising campaign and worked to foster trade relations with other countries in an effort to open new markets. It was active in the defense against the counterfeiters of Havana cigars abroad and created the so-called Official Havana Cigar Centers, the first pilot center being opened in Miami in the forties, with additional centers planned for other locations in the United States, Europe and South America.

This multifaceted activity by the Commission was in many senses quite useful, but it lacked the necessary weight to have any influence or real impact on the industry. However, the Experimental Station at San Juan y Martinez, which started operations in 1937, did play a very important role in future projects through its work to support the development by the industry of the production of cigarettes using blond tobacco, and through its assistance in the growing of the proper leaf, which was brought over from the United States.

In 1940, the Register of Exporters kept by the National Commission for the Advertising and Defense of Havana Cigars contained 40 firms owning 302 recognized brands.

**Vanity rings of H. Upmann. (Same size).**

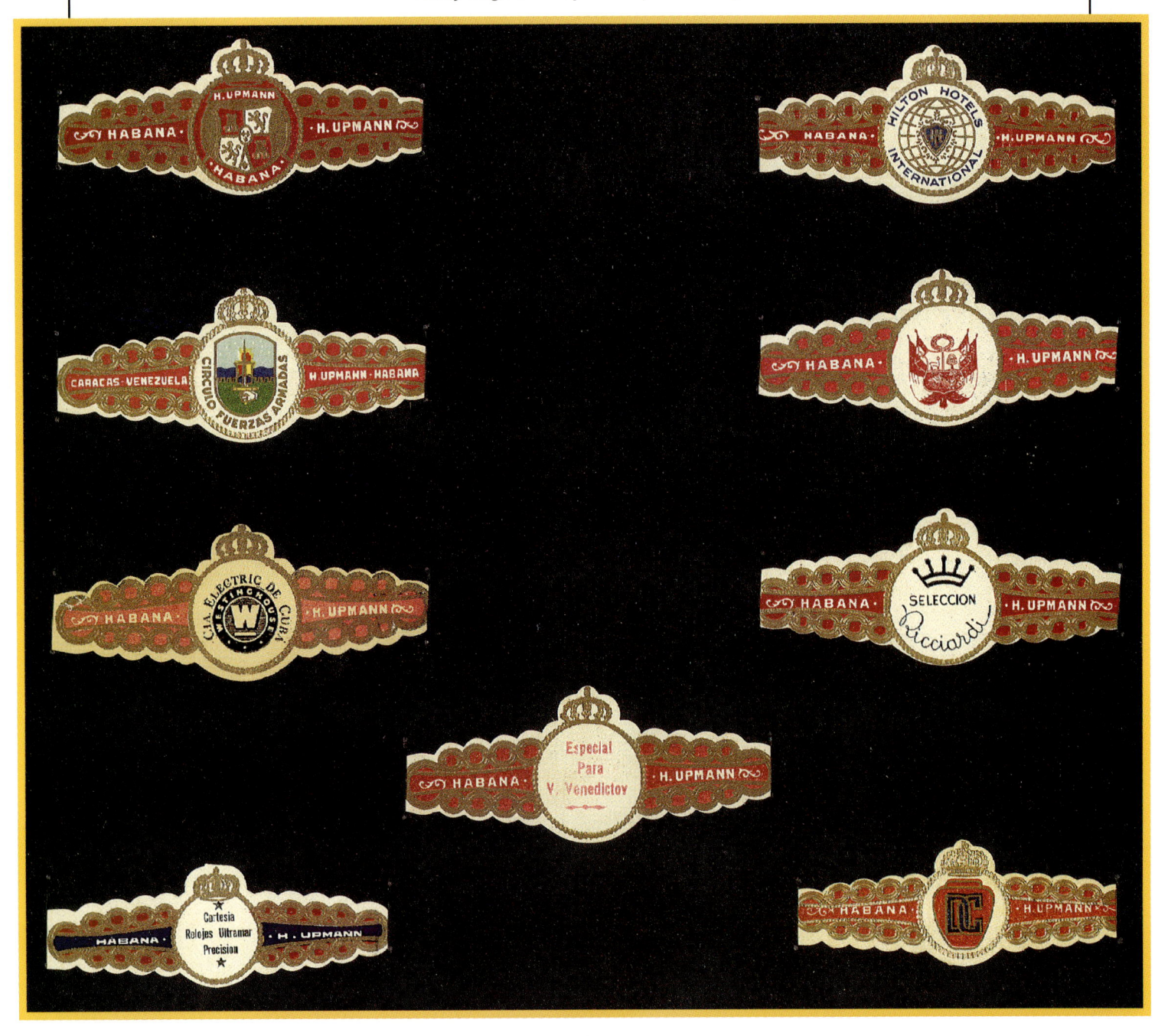

**A large cigar box and a small label done by Manuel Lopez company for their Valle brand.**

In October of 1945, a Special Commission was created to make an immediate study to design a plan for the introduction in Cuba of cigar-making machines. In January of 1946, this Commission resolved to make a Census of Workers of the Cigar Industry, with the total financial support of the National Commission for the Advertising and Defense of Havana Cigars. The census began on January 14 and concluded on September 30, 1946.

The census revealed that between October of 1944 and September of 1945, there were 1,050 cigar manufacturing facilities in the country, which had employed 12,286 cigar makers and 734 cigar ringers. In December of 1945, 8,510 cigar makers and 525 cigar ringers were employed.

The working cigar makers were distributed as follows: Havana (city), 3,300; rest of Havana Province, 1,048; Las Villas Province, 2,152; Pinar del Rio Province, 763; Orient Province, 761; Camaguey Province, 322; and Matanzas Province, 164.

In December of 1946, the Government instituted a measure that was beneficial for the cigar workers. The Gaceta Oficial de la Republica de Cuba published the Law on the Retirement of Cigar Workers, pursuant to which all workers in the cigar industry were given the right to retire and their families or beneficiaries were given the right to a pension.

In 1948, the figure for Havana cigar exports was 53,811,000 units with an overall value of $32,669,314, down just over $2,000,000 from the previous year.

The crisis that began after the war led to a catastrophic loss of markets. One of the consequences of the serious damage sustained by the economies of the countries of Europe from the conflagration was that the purchasing power of many that were traditional consumers of Havana cigars dropped to minimal levels. As a result of the closing of the Spanish market in 1949, only 21,000,000 cigars were exported, of which the United States accounted for half and Spain accounted for 10%. Prices, however, improved toward the end of the decade, rising to $173.31 per thousand in 1949 and $184.69 per thousand in 1950, as compared to the $114.52 per thousand averaged in the years between 1944 and 1948.

The shortage of hard currency and the defense measures implemented by the countries to prevent the flight of the little they had; the necessary priorities set on the acquisition of essential products (which did not include cigars); the endeavors of all countries to protect their own producers; the absence of credit instruments; and the agreements which, to deal with the shortage of convertible currencies created agencies to regulate commercial transactions, all had a profound impact on the Havana cigar trade.

**La Alhambra brand by Manuel Lopez. This is a small box label. (Same size).**

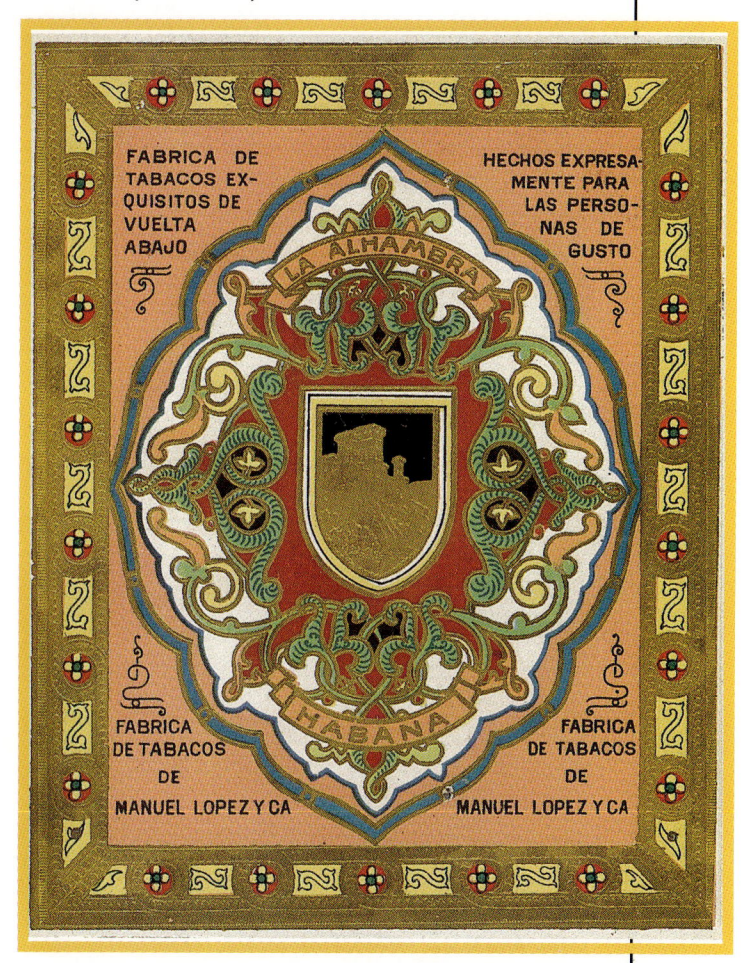

**An intricately designed label for La Alhambra brand by Manuel Lopez. (Same size).**

Thus objective observers of the panorama presented by the Cuban cigar industry in the 50s came to the conclusion that immediate action was necessary in the international sphere to remove the difficulties hampering exports. This goal was completely possible if the potential buyers were offered the opportunity to acquire cigars with deferred payment or through compensation agreements in exchange for agricultural or industrial products.

They were convinced that this expedient could help recover some of the lost markets. The decade that began in 1950 presented the industry with new and complex prospects. The preceding and immediately ensuing years of the post-war period presented a panorama of intense desolation and crisis. The ephemeral recovery of cigar markets in the war years brought a respite, a momentary breath of fresh air for the export manufacturing sector. But once these cyclical moments of "bullish" exports, known in Cuba as the "Fat Cow" years were over, the demand dropped again with the end of the war and the counter cycle ensued, the critical period known as the "Lean Cows."

Coinciding with this period and partly as a result of it, the contradictions between workers and employers became more intense.

The severe problems facing the sector, the drop in exports, the lay-offs, the low salaries and international political situations were all reflected in the taking of radical positions that forged divisions.

The effects of the World War had a lasting effect on the cigar industry. The chaos and the drop in production made it clear that emergency actions were needed to reactivate the industry based on a modernization of the factory process, the reopening of lost markets and the opening of new ones, an aspiration that would be possible for a product that could count on lower prices as one of its competitive advantages, thereby making it accessible to a greater number of consumers.

Beginning in the mid-40s, encouraged by the attitude of the administration in government at the time, several manufacturers decided to introduce the cigar rolling machine. They were hoping to get around the limitations and the terms of the denial that had stopped **Por Larranaga** cold in its tracks when it made the attempt in 1925. The united front that had stood up to and defeated the original attempt was cracking up. There was growing division among factory owners and the differences between them and the workers were getting more and more antagonistic. The position defended by the workers initially was flat out rejection, but the very unions changed the target of their opposition by aiming their concerns and demands not on the machine itself, but on the conditions under which these would be introduced. The 1946

**La Moda Gran Fabrica de Tabacos is the name of this short-lived venture. This label is very rare. (Same size).**

census was one of their demands, as were the introduction of subsidized payment for workers who had been laid off, respect for seniority rights, first option for jobs in the machine department for laid-off workers, and the limitation of mechanized manufacture exclusively to cigars made for export.

In 1950, by virtue of Decree No. 1073, the Cuban Government finally approved the introduction of mechanization. Machines were to produce all cigars for export, plus 20% of cigars for domestic consumption.

Against a somber background of labor conflicts and controversies, the cigar industry was once again rising to the challenge, now with the added momentum of its transformation. The United States and Europe were emerging from their most difficult period following the Second World War and were giving signs of a recovery. The Cuban cigar industry's recovery followed along at more or less the same pace in its offensive to re-open old markets. The first

**Exquisitely designed label for La Flor de Murias, an Antonio Murias brand.**

commercial agreements were signed quite early in the decade (1951) with England and West Germany providing certain facilities for cigar imports. Other agreements were signed with the United States and Spain, although in the United States the high customs duties continued to be an impediment.

The Cuban offensive to conquer foreign markets took full advantage of the activities being promoted on the international scene by United Nations agencies, such as the First Conference on Tariffs and Trade, which began in Torquay, England and concluded in Geneva with the General Agreement on Trade and Tariffs known as GATT. Also important was the World Tobacco Congress held in Amsterdam, Holland, where a Cuban representative was elected to form part of the Committee of the International Tobacco Organization.

**Some labels have no identification or country of origin on them even though it was against the law!**

In the meanwhile, the National Commission for Advertising and Defense of the Havana Cigar assumed the role for which the manufacturers had been demanding a Government agency for almost a Century. The Commission strove to ensure the purity of the Havana cigar and the legal measures that governed the international cigar trade. It also worked to create an awareness of the qualities of the genuine article to promote sales increases and open new markets. As part of its endeavors, it opened the Special Havana Cigar Center in Miami, Florida, and its special representatives in New Orleans, South America and Europe ensured that the Cuban industry was represented at trade fairs and exhibits, such as the ones held during that period in Barcelona, Chicago, New Orleans and Osaka.

It also organized contests and established a series of awards that were given to a number of institutions, political personalities, scientists, intellectuals, athletes and artists around the world.

The confluence of these joint efforts to promote the cigar industry tended to compensate for the neglect it had suffered in the past and to provide Havana cigars with on-going and multifaceted protection in international trade.

The renovation of relations with England produced a very encouraging return of Havana cigars to one of its most important markets. The bottom had fallen out of the English market with the outbreak of war in 1939, precisely at a time when it was the number one market for Cuban cigars. In the years immediately following 1945, the year of victory in Europe and peace, there was nothing that could be done in that market because England had suffered the devastating effects of the war and in its resulting economic anemia was carefully channelling every last bit of its scarce hard currency into priority purchases. Its recovery, although still rather slow, did allow both sides to make their interests more compatible and on August 10, 1951, a new treaty was signed, which was to be in force as of March, 1952, and which provided for purchases of Havana cigars for a value of $500,000 a year.

The decade had many important repercussions in the industry's relations with Spain. The old treaty signed in 1927 was replaced by the Commercial and Payments Treaty of 1951 and exports of Cuban cigars to Spain took a quantum leap. From just over 2,000,000 cigars in 1950, sales soared to almost 14,000,000 in 1951. In a relatively short time, Spain became the most important importer of Havana cigars. In fact, there were years when it accounted for almost 50% of total Havana cigar exports. This was the case in 1957, when the Spanish market consumed 36,242,000 cigars.

The marketing offensive targeting the United States went through a set of special circumstances. Negotiations there still had to deal with the long-standing duties burden. The existence of GATT, under the auspices of the United Nations, created a certain optimism regarding the possibility for better understanding and trade relations among countries. But the Cuban cigar industry felt that it had benefitted very little from those mechanisms, especially in markets that were of capital im-

**Vanity bands of Maria Guerrero. (Same size).**

**Martinez Brothers of Havana produced this modernistic label for their brand Los Statos DELUXE.**

portance for Cuban products. In the United States, the old protectionist legislation was still a factor, so efforts had to be maintained to try to achieve a review of the tariffs and arrive at a new agreement.

Despite the tariffs, however, consumption of Cuban cigars increased vigorously in the United States in those very same years (the 50s), most likely due to the production of less expensive and more competitive cigars, plus the promotional work of the Special Centers, better marketing and sales techniques, increased advertising and publicity.

In 1950, a slow year for the industry with exports of just over 20,000,000 units, the United States took about 11,500,000. In the ensuing years, US imports of Cuban cigars continued to grow until they reached their plateau, generally below Spain but very significant.

In 1956, the reduction of customs duties on cigars imported from Cuba, something the United States had been promising the island for some time, finally became a reality. The reduction would amount to 15% on the basis of 5% reductions each year over a period of three years, which was the limit imposed by existing U.S. legislation. Cuban cigars had to pay duties of $1.50 per pound, plus 10% ad valorem. The reductions would work the duties down step by step. The first cut would bring the duty down to $1.42, plus 9.5%; the second would take it to $1.35 and 9.0%; and the third would leave it at $1.27, plus 8.5%.

Canada followed the United States in reducing the duties on Cuban cigars, but did it in a single stroke bringing them down from $1.75 per pound and 15% ad valorem, to $1.50 and 10%, respectively.

Spain, the United States, the United Kingdom and France became the leading markets for Havana cigars, although the expansion of the Cuban marketing drive reached the tastes of smokers all over the world. Many other countries, including Uruguay, Canary Islands, Sweden, Switzerland, Germany, Jamaica, Peru, Egypt, Iceland, Portugal, Argentina, Norway, Austria, Chile, Venezuela, Canada, English Africa, Japan, Mexico, St.Pierre and Miquelon, Hungary, Luxemburg, Yugoslavia, Fernando Po, North Rhodesia, Tunisia, the Union of South Africa, Belgium, Panama, Hong Kong and so on, up to the very significant figure of 75 markets, participated in the demand.

After the first half of the decade, the National

De Pino, Villamil y Ca. produced this lovely label for its brand La Africana, which, unfortunately, has disappeared from the market. (Enlarged 40%).

*This Los Statos Deluxe label appeared in many forms.*

Commission for Advertising and Defense of Havana Cigars was convinced that the mechanization of the export product, management, trade treaties, careful attention to the markets and prompt compliance with purchase orders were responsible for the recovery.

In 1954, a year when there was a slight drop in exports, it was estimated that there were 1,050 cigar factories in the country and some 527 private producers. Of the former, 91 were producing for export and domestic consumption simultaneously, while the remaining 909 and the private manufacturers were producing for the domestic market.

The increasingly significant presence of Havana cigars in the markets of the world unleashed another wave of imitations and counterfeits. Among the most notorious were manufacturers in England, Germany, The United States, Canary Islands, Brazil, Belgium, Chile, Argentina, Paraguay and Morocco. The Defense Commission acted more or less successfully within the juridical constraints of each of these countries. A much publicized case was the ilegal use of the phrase "Havana Tobacco" on cigars manufactured by a British firm. They were was forced to add that the cigars were "Rolled in England." A more rigorous claim was filed against a manufacturer in Paraguay who had falsified the lithograph of the famous **Hoyo de Monterrey** brand. These and other cases were based on a demand for respect for international law and the various conventions on the protection of brands and industrial property, as well as the use of geographic denominations in the claims made on their origin. The guarantee seal, which had been established in Cuba by law in 1912, constituted a brand belonging to the Cuban State and recorded in the corresponding register.

This Lords of England label appears in several sizes and is cropped (cut) into the sizes necessary for the purpose. (Enlarged 30%).

# Chapter Twelve
# EXPORTS DURING THE FIFTIES

Cuban cigars began their take-off in the very first years of the decade. The export figure in 1950 was a mere 20,424,000 units. In 1952 it had risen to 36,956,000 and in 1952 it remained almost steady at 37,635,000. Then in 1954 it shot up again to some 44,875,070 and slipped back to 41,673,000 in 1954. In 1955 there was another leap to 52,869,000 units, followed by yet another leap in 1956 to 62,554,000. The year 1957 turned in one of the best performances since 1930, with a total of 73,496,000, surpassed only by the figures for the war years of 1944 and 1945. That year it was evident that purchases by countries with which there had been bilateral negotiations accounted for 89.8% of the overall export volume. Quite significantly, Argentina, which was not a great importer, did however increase its purchases by 90% over 1956.

In 1958 the upward trend peaked at 79,878,449 units. A large part of the increase was due to the high demand mainly in Spain, the United States, the United Kingdom and France.

The aggregate value of yearly exports are given in the following table:
1950 $2,139,910
1951 $6,186,000
1952 $6,551,696
1953 $7,365,582
1954 $6,996,000
1955 $8,776,372
1956 $9,901,165
1957 $11,128,000
1958 $12,255,552

In a business world characterized by tough competition in generic and brand products, the Havana cigar was able to recover its dominant position in the market by never letting up in the fight.

It was armed, however, with its most prominent weapon: its unblemished quality and the preference that expert smokers have at all times reserved for the enrichment of tradition.

The Havana cigar is a precious jewel in the world of smokers, and a heraldic mark of the Cuban nation. But it is also a symbol of deep-rooted patriotism, for the order sent by Jose Marti in 1895 from the United States to the patriots in Cuba to begin the revolutionary war was hidden inside the filler of a cigar. This is why February 24, the commemoration of the revolutionary pronouncement at Baire, which began the war for independence, has also been appointed the Day of the Cuban Cigar.

At the World Conference on Commerce and Employment, which began in Geneva in 1947 and concluded in Havana in 1948, during the debates on the legitimacy of origin of the different products of each country, the countries that signed the Multilateral Agreement all agreed that Havana Cigars were "solely and exclusively those produced and manufactured in Cuba."

Toward the end of the 50s, the concentration

**A label for the size of the La Preferencia box. (Same size).**

of the industry in the hands of a few companies, relocation of a number of manufacturers some time later, and the flight of capital, brought a certain intentional decline in production in an attempt to increase prices and destabilize the industry. This situation affected the domestic demand and endangered the supply of markets where Havana cigars were traditionally available. The ensuing wave of unemployment that hit the workers again led to protests by the men and women of the sector. The manufacturers, fearful of the political changes, withdrew to damage the national economy.

This was the economic and social panorama the Revolutionary Government had to deal with in the cigar industry. Thus on September 15, 1960, Resolution 20260 issued by the Ministry of Labor decreed the intervention of all the cigar and cigarette factories and all the tobacco warehouses and deposits in the country.

As of that moment, cigar production and supplies to the international markets of the coveted Havana cigar began to recover stability and to this date they continue to reach knowledgeable smokers the world over.

# Chapter Thirteen
# FACTORIES AND BRANDS

The historical outline of the Cuban cigar industry presented thus far, cannot be summed up without reviewing some of the main Havana cigar factories that have manufactured and exported their prized vitolas with meticulous care and invariable quality for more than 150 years.

The story of these factories is inextricably linked to their founders, the men who wisely and painstakingly turned tobacco into the nation's second ranking industry and promoted the names of Cuba and Havana cigars in the most distant corners of the Earth.

As a fitting and proper tribute, this chapter offers a description of some of the best known factories and cigar brands that are still exported, as well as facts, figures and circumstances that are part and parcel of their respective histories.

## PARTAGAS

In 1867, at the Paris World Fair, a brand of Havana cigars won the coveted Gold Medal for its outstanding quality. That great success of the Cuban cigar industry was to be scored again at the same venue in 1878 and 1889, as well as in other fairs and exhibitions where the makers of Partagas cigars chose to be present. Since then, the golden award became a distinctive feature of the Partagas label signifying the standard of quality of a product that had come to be the choice of discriminating smokers the world over.

By the year 1861, the Real Fabrica de Tabacos Partagas already held a prominent position. It had contributed like no other cigar maker to the conquest of the biggest and richest markets in the world by the quality of its vitolas, made with the best tobacco leaves from the Vuelta Abajo region.

The founder, Jaime Partagas Ravelo, was a daring and enthusiastic Catalonian, full of energy, resolve and ambition, gave his name to the factory in 1845. Partagas belonged to the sector of free cigar manufacturers and by 1827 he was already running his own small shop. A few years later he bought a large portion of the Hato de la Cruz hacienda, in the municipality of Consolacion del Sur, province of Pinar del Rio province. In time he came to own several tobacco plantations in the land where the prized Vuelta Abajo leaves are grown. Always a sharp businessman, Partagas made it a point to control the entire process, from the fields to the final product.

By owning some of the best tobacco lands in Pinar del Rio, he was able to oversee directly each and every detail pertaining to his valuable raw material, and thus guarantee a steady supply of only the best leaves to his factory. National and foreign smokers came to enjoy and admire the excellence of Partagas cigars and made them their choice. The well deserved fame of Partagas became a legend. In trying to explain the matchless quality of the produce, rumors pointed to secret, arcane procedures known only to the prosperous cigar-maker. There was also speculation on the origin of the leaves and on the process of fermentation or maturing of the stripped tobacco whereby leaves were stored one by one in barrels hung in "barbacoas" and left to mature. For some kinds of leaves this process took between four and twelve months; for others it could take up as much as six or seven years.

Jaime Partagas knew that the secret lied there, in the barns where the aroma of tobacco leaves, while fermenting and aging, becomes blended and unified in much the same way as do the spirits from the grapes in wine cellars. A

**A photo of the Partagas factory which still exists as it was 93 years ago!**

Jaime Partagas, the founder of the original factory which still exists and is operational in Havana.

An old, black label with gold embossing and stamping, making this a very elegant but inexpensive label to produce.

careful processing, rigorous storage procedures and selective aging periods of the tobacco leaves constituted the real secret of Partagas, who was always mindful of having great quantities of tobacco in each of those phases.

Another factor that determined the particular taste and fragrance of each make and their invariable bouquet was the mixture of different kinds or classes of leaves with varying degrees of aging, which became typical for each vitola. At certain moments, Partagas applied an experimental method based mainly on trial and error, and on the accumulated experience of harvesters, blenders, factory operators and smokers' preferences to achieve blends that were fully capable of satisfying even the most demanding clients.

As a result of his ceaseless research and creativity, Partagas developed 67 different vitolas or makes like Carolinas, Cosacos, Coronas, Coronas Grandes, Cervantes, Cazadores, Delicados, Entreactos, Franciscanos, Hermosos No.4, Julietas No.2, Londres, Minutos, Carlotas, Marevas, Ninfas, Partagas Nos. 3, 2 and 1, Panetelas, Superiores, Britanicos, Piramides, Standard, Nacionales, Cremas, Petit Cetros, Belvederes, and many others.

With a keen eye for business and ready to incorporate swiftly any profitable change in the industry, Partagas brought into his factory the new forms of fancy packaging and publicity conceived and introduced by Ramon Allones, elements that were rapidly becoming essential in the increasingly competitive world of cigar manufacturers. In his plans for advancement and modernization, Partagas treated his workers with considerable open-mindedness. He immediately accepted the request presented by the Cigar Workers Association of Havana to introduce reading in his shops. This idea had been advanced by the Asturian Saturnino Martinez, a cigar roller at the Partagas factory who was also the editor of the workers' newsletter "La Aurora," where he defended the adoption of this procedure to enhance the moral and intellectual level of his fellow workers.

On 6 January 1866, reading out loud for the cigar workers was performed for the first time at the Partagas factory. Don Jaime had spoken highly of the idea and consented on one condition:that all materials to be read had to be previously presented to him personally for his approval. Partagas himself was present on the inaugural day and offered to erect a lectern at the center of the workshop's gallery so that everyone could hear the voice of the reader clearly.

Due to the credit gained by the brand Partagas, many outstanding personalities, eager to see for themselves how the famous cigars were made, came to visit the factory. Noteworthy foreign guests like the US Secretary of State, Mr. William H. Seward and his son, toured the factory on 22 January 1866 and walked into the large gallery where the workers rolled cigars while listening to the reader deliver "El Rey del Mundo," by the popular Spanish novelist Manuel Fernandez y Gonzalez. Impressed by the industriousness and attention of the craftsmen, the visitor expressed his admiration and congratulated Mr. Partagas and his workers.

Other visitors in those days also had

**An early Partagas name card. (Same size).**

favourable comments regarding the new procedure. On 3 February that same year, Partagas fulfilled his offer by inaugurating the reader's lectern, the first of many to be erected in almost every cigar factory in Cuba. The donor delivered a short speech after presenting the lectern and one of the workers read words of gratitude on behalf of all his colleagues.

Since the mid nineteenth century, a growing list of consumers from all parts of the world manifested their preference for the products of Partagas. Demand grew significantly in Spain. the United States, England, Germany, Holland, Belgium, Italy, France, Switzerland, Russia, Egypt, Rhodesia, Canada, Argentina, Mexico and other countries. This impressive array notwithstanding, Partagas kept expanding the range of his products and continued to search for new clients, be them for his already famous Havanas or for the cigarettes and pipe tobacco that he also manufactured.

A modern Partagas box label with the wooden grain printed on paper.

At the peak of his entrepeneurial accomplishments and still in the prime of his life, Don Jaime Partagas, the idefatigable tobacco industrialist, met a sudden death. On the night of 17 June 1868, on the way back to his home at the Vista Alegre farm after visiting some friends, he received gunshot wounds that caused him his death one month later, on 17 July.

After his abrupt demise there was a brief period of uncertainty regarding the future of the emporium built and steered by Partagas based on his enterprising nature and iron will-power. The heirs of Don Jaime remained in business for a number of years using the same instruments and methods created by the founder, but finally sold the factory and the lands. Jose A. Bances, a wealthy banker who also included tobacco among his many trades, became the proprietor in the Nineties, when business began to ebb.

In 1899 Bances joined in partnership with and later sold his share of the factory to Mr. Ramon Cifuentes Llano, who already owned a tobacco warehouse in Amistad street, in Havana, as well as a cigar factory in association with a Mr. B. Fernandez in 160 Industria street, very near the Partagas industry. Cifuentes proved to be as capable and ingenious a businessman as the late Don Jaime Partagas. Upon concluding the severance arragements with Bances, he went into full partnership with Mr Fernandez and founded Cifuentes, Fernandez y Cia. owners of the Partagas industry. This marked the begining of a new stage of development for the brand of Partagas.

This entity was dissolved in 1916 and Cifuentes took up Mr. Francisco Pego Pita as a partner to create the firm Cifuentes, Pego y Cia. an association that would last until 1941. Cifuentes managed to raise the excellence of Partagas products to even higher standards and thereby consolidate the firm's financial health. He bought and incorporated several well established cigar brands like Ramon Allones, F. Pego Pita, La Intimidad, Coruncho, La Eminencia, El Corojo, Particulares, Guayarre, Prudencio Rabell, La Insuperable and El Cambio Real among others. But perhaps the most important contribution made by Ramon Cifuentes to the brand of Partagas was the meticulous training that he gave his sons and heirs on cigar industry management and the handing down of the "secrets" and know-how of the trade.

Initially, the Partagas factory was located at the corner of Industria and Barcelona streets in Havana. The original building, made of bricks and wood, no longer exists. Its second location was a house marked 158, at Industria street

**The Flor de Tabacos cigar box end by Partagas.**

**Ramon Cifuentes Llano.**

**Lightly imprinted on the reverse side of many Partagas labels is a sketch of the original structure of the Partagas factory.**

**The Partagas factory faces the Capitolio in Havana.**

before finally occupying the facilities formerly marked with the numbers 170, 172 and 174 on the west side of that same street. The factory remains there today, marked with the number 520, between the streets Dragones and Barcelona. The original factory was located right in front of the old Villanueva railway station, which was very practical for the swift

**The present (1996) Partagas factory.**

102

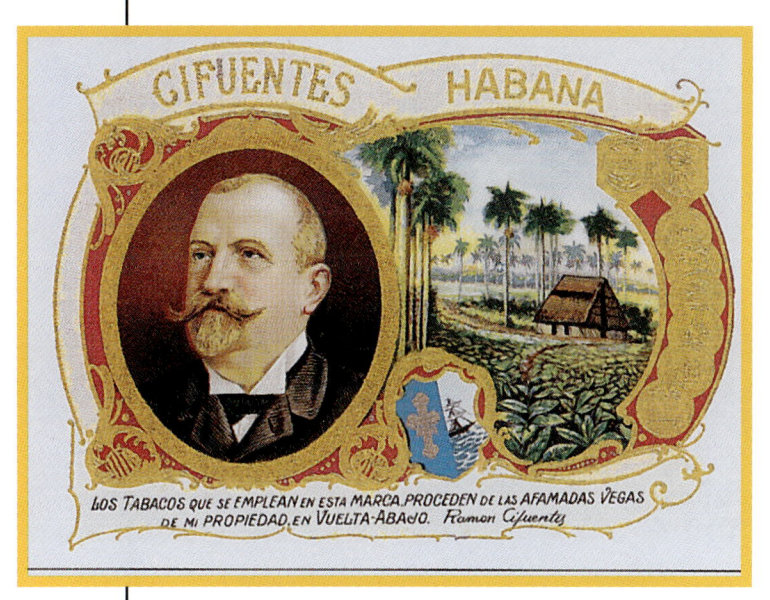

**Ramon Cifuentes issued this vanity label.**

transportation of the plant's supplies and output. The station was brought down in the Twenties to erect in its grounds the majestic Capitolio building—an exact duplicate of the Capitol in Washington D.C.—where the Cuban Legislature sat from 1929 to 1959. Today the Capitolio houses the Ministry of Science, Technology and Environment. The Partagas building is made out of concrete, blocks and bricks with iron columns and occupies an area of 1,369 square meters. Three colonial style stories

**Another view of the present Partagas factory. It hasn't changed since it was built.**

Inside the showroom in the Partagas factory is a special room where exceptionally fine cigars of all labels are offered to special clients. The special guests are on the right and left hand sides, while the three gentlemen are managers of the factory.

103

*The magnificent label produced to celebrate another gold medal.*

surround a beautiful central patio on the ground floor. These higher stories form an independent section thst is linked, nevertheless, to the adjoining two storied facility. The plant's main facade was extensively remodelled to harmonize its lines with the classic design of the Capitolio.

Ramon Cifuentes died in Spain in April 1938 and his partner Francisco Pego Pita passed away in Havana, in November 1940. The heirs of Pego Pita were disassociated from the firm, which left Cifuentes' relatives, Ramon, Rafael and Manuel Cifuentes Toriello and Leandro Cifuentes Alvarez, as sole owners. The first three took up the factory's management.

Partagas' operations were not always a bed of roses. There were times of economic depression as a result of national and international junctures. These fluctuations caused tensions and instability affecting both the workers and production with the consequent upsurge of labor conflicts.

One of the first ocurred in 1857, after the United States erected tariff barriers for all Cuban cigars and tobacco leaf imports. The first and second wars of independence (1868-1878 and 1895-1898) brought turmoil and depression for the industry. The impact of the Mac Kinley Bill between 1891 and 1893 has already been discussed.

For Partagas the hard times meant a decline in its production levels until the turn of the century, when a steady recovery brought the factory back to its usual standards of output.

The 1902 and 1907 strikes in defense of the admission of native personnel to better paid jobs and for payment of their salaries in US dollars, as well as the 1930 and 1932 strikes against wage cuts in the industry were all backed by the workers of Partagas. In view of the outcome of these last strikes, the owners decided to open a smaller factory in Bejucal, a town near Havana, where wages were lower. The main factory in Industria St. was kept running with a small staff of skilled workers that turned out only minimal production aimed at the tourist market. When the conflict was finally settled in 1939 after the signing of new labor contracts with the cigar workers organizations, the administration returned to its traditional factory in Havana and resumed its normal output rates.

The closing of the European markets during the First World War from 1914 to 1918, as well as the severe crisis caused by the bank and market crashes in 1920 and 1929 struck Partagas just as hardly as any other name in the industry, but the firm overcame the throes and endured.

During the Second World War from 1941 to 1945, the managers of Partagas opened a new factory at the corner of Figuras and Lealtad streets to meet the soaring demand of cigars that started coming from the United States. The boom ended abruptly after the war; orders were cancelled, production ebbed and massive lay-offs determined the closing down of the new factory, which generated labor conflicts until 1947. In the early fifties, Partagas began introducing machines into the cigar manufacturing process. This naturally meant great increase in productivity but also considerable staff reductions.

Shocked by their displacement by machines, the workers initially expressed their rejection but finally had to give in to automation. A clear distinction was then established: export cigars would continue to be rolled manually, cast rolled and by the electric machines. The latter two kinds would be manufactured with either short or long filler leafs. Many outstanding personalities who were also cigar aficionados became regular clients of Partagas. In his ode "The Betrothed," Rudyard Kipling, the famous British author who was the Nobel Laureate for Literature in 1906, candidly declared: "I have been a priesty of Partagas a matter of seven years."

Among the notable Heads of State who were dedicated fans of Partagas one finds the Duke of Windsor, Prince Olaf of Norway, Prince Frederick of Denmark, King Farouk of Egypt, Prince Raniero of Monaco and pointedly Sir Winston Churchill, the British Prime Minister during World War II who on one occassion, when asked about the island, he answered "I always carry Cuba on my lips."

In the past and the present centuries prominent men of letters and performing artist like Honore de Balzac, Victor Hugo, Alfred de Musset, Franz Liszt, Mark Twain, Somerset Maugham, Ernest Hemingway, Orson Wells and John Wayne enjoyed the cigars by Partagas. In the world of finance and big business, the

**Inside the special smoking room on the ground level of the Partagas factory, many brands are offered. All of these brands and sizes are made at the Partagas factory, most of which do not carry the Partagas label.**

The familiar Partagas box label.

## Menú

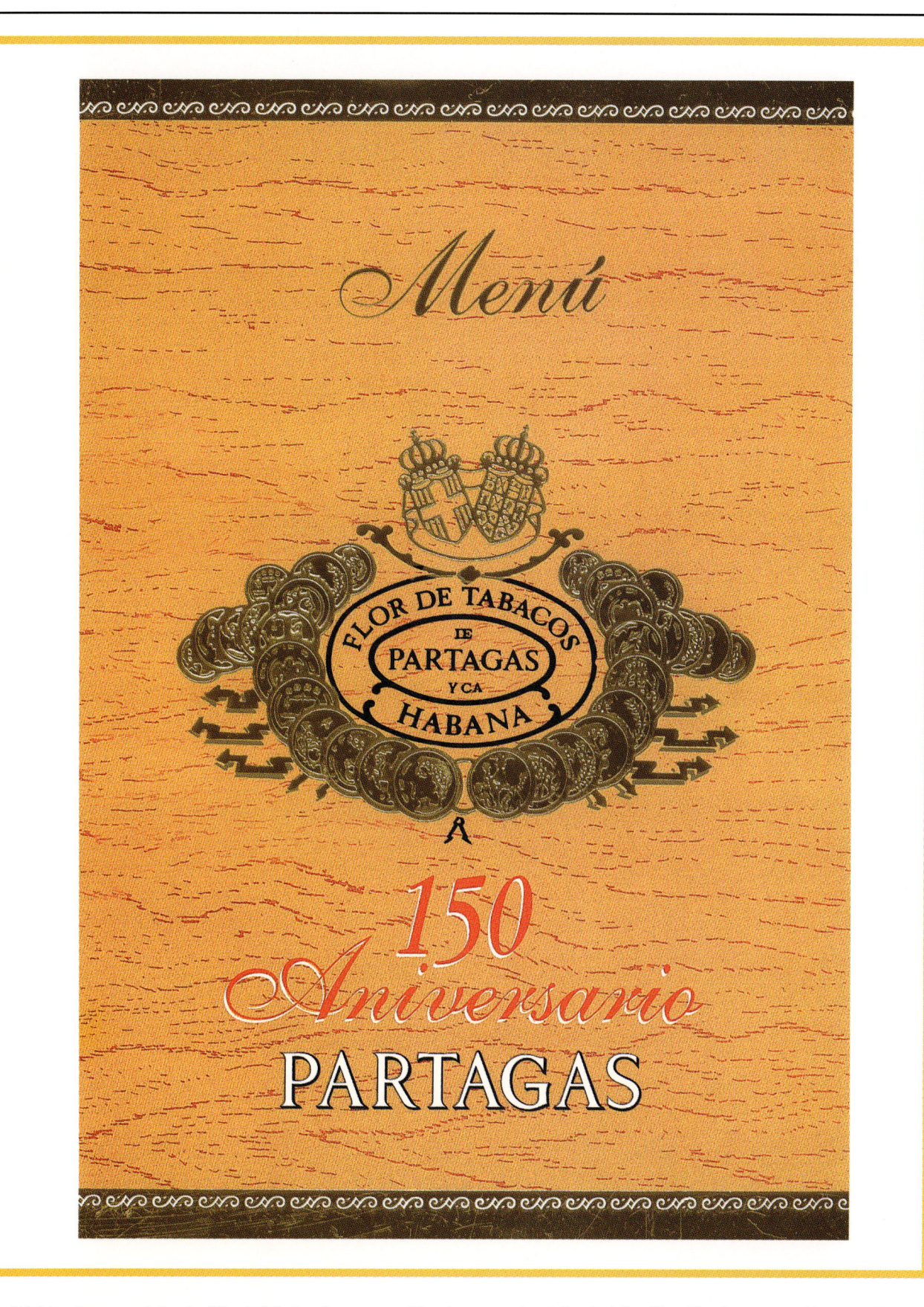

### FLOR DE TABACOS DE PARTAGAS Y CA HABANA

## 150 Aniversario PARTAGAS

In 1996 Partagas celebrated its 150th Anniversary with a huge party at the Cohiba Hotel in downtown Havana. Everybody of tobacco fame was there. Special cigars and champagnes were served to the guests. The cigars had special bands (see to the right)

names of Dupont de Nemours, the Baron of Rothschild, Alfred Dunhill and Aristotle Onassis are but a few among the smokers of Partagas, and famous military like Marshall Ferdinand Foch and Colonel Sosthenes Behn are also members of this select club. Partagas supplied its products to the Royal Houses of Spain and Italy and received orders from the farthest corners of the Earth.

In August 1939, Mr. J.F.Vernet, a large importer of champagne and brandy in Dairen, Manchuria, asked to contact a first-class exporting firm of Havana cigars to purchase the product for distribution not only in Manchuria, but in the entire North of China. The request was received and accepted by Partagas.

As a capitalist entity, the Partagas administration established insurance policies against accidents, load elevator hazards and fires within the premises of the main factory and all its warehouses. Their export products were also covered by insurance until arrival at their points of destination. Insurance covering all properties of Partagas in 1958 amounted to 1.4 million pesos.

At the moment of its intervention in September 1960, it was factually established that Partagas and five other tobacco firms controlled 97 per cent of Cuban cigar exports, a remarkable business achievement that was envisaged and pursued by Don Jaime Partagas Ravelo as far back as 1827.

**Sir Winston Churchill receiving a box of Habanos from the Minister of Agriculture in Havana. Churchill was an avid cigar smoker and many large size cigars used the name Churchill to designate the size.**

**At the 150th Anniversary of Partagas special guests were treated royally. The gentleman in the center is the manager of the Hotel Cohiba and is smoking Cuba's largest boxed cigar, a Diadema. Each Diadema comes in its own separate box.**

# H. UPMANN

By the end of the first half of the 19th Century, just when the Cuban cigar industry was begining to bloom, the firm H. Upmann inaugurated its operations in the island. There was no way of telling then that the name Upmann would become the trademark of one of the largest and most prestigious manufacturers of Havana cigars, cherished and sought after by the most dicriminating smokers the world over.

The Upmann brothers, Herman and August, born and bred in Bremen, Germany, arrived in Cuba in 1843 looking for business oportunities to which they could dedicate their energy, savings and talent. On 1st May 1844 they opened a store in Havana and became established as general merchants. From the very outset the firm's chief trade was the manufacture of high quality hand-rolled cigars. Small banking operations became their second priority. Twentyfour years later the Upmann brothers were solidly established as bankers while their cigar factory already enjoyed international fame. The firm H. Upmann was to be one of Cuba's leading banks until its foreclosure in 1922.

The investment made by the Upmanns was the first direct input of German capital in the Cuban cigar industry. The initial factory was called "La Madama," but this name slowly faded away and only the veteran workers used it to identify the new factory inaugurated in 1890. Worldwide recognition went to the brand registered by the firm: H. Upmann.

Early on, the brand secured a share of the growing demand of Havana cigars in both the national and international markets. The exquisite aroma, fine flavor and superb finish of the H. Upmann vitolas—made exclusively with choice leaves from the Vuelta Abajo plantations—soon won the preference of consumers who praised

**The H. Upmann family in 1921**

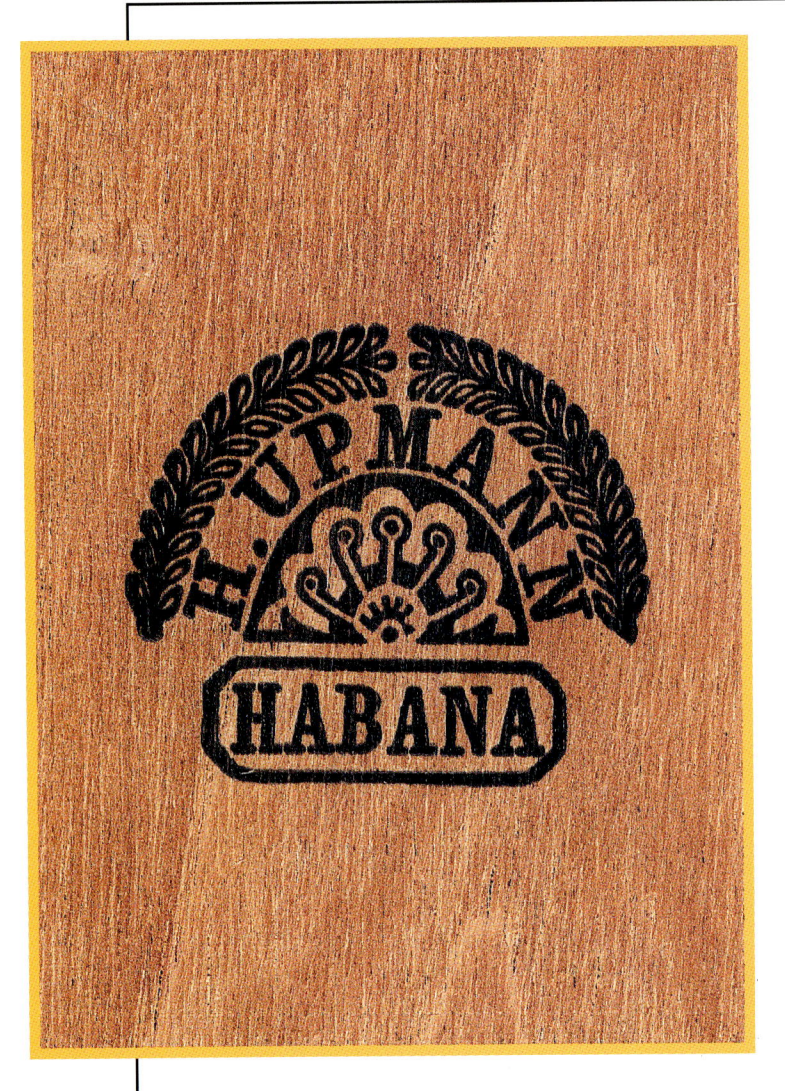

The Upmann brand printed on cedar. (Same size).

their excellence. In time, while the industry grew and the rolling, wrapping and packaging of Havanas became more sophisticated, the H. Upmann firm experienced a an ever growing demand for their produce in every market in the world and by the most outstanding international celebrities. By then the Upmanns had joined in partnership with Heinrich Kaufsen, a fellow German, to open a second factory, right beside the first, named "Flor del Pacifico." The boom was so great that the initial factory in 75 San Miguel St. proved too small to meet the demand. In 1886 the firm incorporated the brand "La Anita" and began to build new industrial facilities more in keeping with the demand and capable of holding its more than 200 workers. The new factory of the very respectable H. Upmann firm, located at 159 Carlos III St. was inaugurated in 1890, but its old workers continued calling it "La Madama."

Herman Upmann died in Bremen in 1894. From his marriage to Marie Braesecke, also Bremen born, came three sons: Herman Albert, Albert Heinrich and Carl Julius. After the passing of the father, his sons Herman—born in 1879—and Albert were sent to Cuba in 1897 and at a very early age began to learn their father's and uncle August's trade. Years later, Carl Julius opened a cigar factory in New York using leaves imported from Cuba but these did not carry the H. Upmann brand which belonged exclusively to his brothers' firm.

When August judged his nephews sufficiently well trained, he left them in charge of the business in Cuba and went back to his native Bremen from where he gave them counsel as senior partner and head of the firm until his death. Herman remained as executive manager of both the bank and the cigar factory, now under H. Upmann and Co.

At the factory, Herman and Albert had the valuable assistance of Paul Meier, a German, and Francisco Fernandez, a Spaniard, both of whom were highly skilled technicians in the art of cigar making and were consequently put in charge of quality control as overseers.

During the depression that hit the tobacco industry at the end of the century British and American consortiums unsuccessfully tried to buy the factory for one million US dollars, plus the estimated price of H. Upmann's facilities and stocks. The initial years of the 20the Century brought a new boom in cigar exports which required the opening of a subsidiary H. Upmann factory in the town of Calabazar, 20 kilometers from Havana. The main plant in Carlos III street again proved too small to hold the 1200 workers—900 men and 300 women—that were needed to meet the soaring purchase orders. H. Upmann's record output for year topped 25 million cigars in more than 200 different sizes and shapes, at prices that ranged from 40 to 150 pesos per thousand units. In November 1907 this steady growth in demand made the Upmann brothers buy the brand "La Flor del Figaro" and in December 1911 the brand "C.G.and Company."

The boom lasted until the outbreak of the First World War; the European conflict imposed considerable constraints to the H. Upmann firm. Since the owners were German subjects, the firm was "black-listed." This meant a house arrest sentence for Herman Upmann and the suspension of all operations for the Upmann's bank until the 1st of May 1920. That was the year of the Great Market Crash that forced many Cuban banks to cancel their payments in the month of October. While others folded under pressure, the Upmann bank did not observe the moratorium decreed on 10 October 1920, a "beau geste" that spoke highly of the bank's respectability earned over long years of operations.

**The old Upmann box label.**

Notwithstanding, when the bank was finally forced to suspend payments in May 1922, the Liquidation Commission that had been officially appointed by the Decree of 21 January 1921, verified that the Upmann brothers were practically bankrupt.

On 30 August 1922, Herman Albert and Albert Heinrich Upmann renewed the registration of their brand of cigars at the Trade Marks and Patent Office, apparently trying to save the H. Upmann brand fron imminent disaster. The eleventh hour effort failed; all properties and goods of the firm H. Upmann and Co. were liquidated and auctioned. The prestigious H. Upmann factory went for only 30,000 pesos. It was the end of the Upmann brothers business.

Herman Upmann died in Havana on 3 September 1925 while the process of liquidation of his firm still continued. Albert married an American and moved to the United States. On 18 December 1922, the Havana branch of the London based Frankau Ltd. a tobacco importing company, took legal possession of the H. Upmann cigar factory. Very soon, in 1924, it was leased to Solaun and Bros. a firm jointly held by Manuel, Bernardino, Francisco and Jose Solaun Gonzalez. These hired the Germans Otto Braddes and Paul Meier, as well as the Spaniard Francisco Fernandez, former specialists who had worked at the H. Upmann factory for more than 24 years, in an effort to revitalize the brand.

The Solaun brothers had come from Spain in 1898 and settled in Puerta de Golpe, at the center of the tobacco-rich province of Pinar del Rio. A few years later they moved to Havana and bought the little known "Baire" cigar factory, located at 34 Belascoain St. Their knowledge of the tobacco trade was vast, but their financial resources were limited. For this reason, when they leased the H. Upmann factory they had to subdivide the facilities of the great Carlos III plant into smaller workshops operated by fewer workers. The first of these shops was set up right at the corner of Belascoain and San Rafael streets where it operated between 1924 and 1928. From there it moved to Figuras St, between Campanario and Lealtad, where it remained until 1937.

Nevertheless, in 1932 due to unresolved labor conflicts, the firm opened a new workshop in the nearby town of Bejucal, while the Calabazar branch continued in operation.

Those were hard years indeed; money was scarce and business fared with uncertainty. Frankau Ltd cancelled the contract with Solaun & Bros. in 1936 and the following year sold the brand over to Menendez y Garcia, who were also of Spanish origin. These partners had a long experience in the industry. They owned the cigar factory "Particulares," in Virtudes St. between Gervasio and Escobar, where they moved the H. Upmann brand, and also a leaf tobacco warehouse in 407–409 Amistad St. between Barcelona and Dragones.

The new partners dedicated all their resources to the task of reviving H. Upmann and among their first measures was the launching of the Montecristo brand. The unique quality of the leaves used in its blend and the guarantee of excellence that the name H.Upmann inspired turned the Montecristos into the driving force of the enterprise and quickly captured the attention of smokers and tradesmen in Cuba and abroad. A surprising sales boom ensued and again a bigger factory was needed in view of the steady flow of ever larger orders that started coming in during the Second World War. This showed that when cigars were made with first-class materials, rolled lovingly by expert hands and carefully packaged into luxurious boxes, an avid market was always secure. In a remarkably short time H. Upmann recovered its standing among the leading manufacturers of Havana cigars producing only the brands H. Upmann, Montecristo and El Patio, because the Particulares was sold to Cifuentes and Co.

The H. Upmann factory was run by a General Manager and each department had a director and an overseer. Until 1934 the work shifts lasted between 10 to 12 hours; then the workers union achieved the 8 hour day and in 1941 the 44 hour week with 48 wage was enforced. Wages were paid by piece and quality standards were strict. Labor discipline was based on the individual worker's need to meet his norm with the required quality. On October 1944, to commemorate its centennial, the H. Upmann brand inaugurated a new factory. Menendez y Garcia erected a huge building at 407 Amistad St. between Barcelona and Dragones that has since been the headquarters of this famous cigar factory.

Many have been the awards and prizes won by the H. Upmann cigars. By special appointment, the brand was the supplier to the Royal House of Spain and won gold medal awards in every international fair it attended: Paris 1855, London 1862, Oporto 1866, Paris 1869, Moscow 1872, Vienna 1873, Chicago 1893, Antwerp 1894, Sydney 1905 and Liege 1907.

**The Solain brothers, the owners of H. Upmann until Menendez bought the factory. This photo was taken in Pinar del Rio at their tobacco plantation.**

The famous Upmann label with the signature and English text such as *This is my signaturë*.

A complete small cardboard box, opened to show all panels, held 5 petite size cigars. (Same size).

113

Its main markets have been England, The United States, Germany, Argentina, Chile, Spain, South Africa, Australia, New Zeland, Canada, Russia, Switzerland, Belgium, Holland, France, Austria, Hungary, Norway and Sweeden.

Depending on the market's demands, the total labor force at H. Upmann, both permanent and part-time, was between 500 and 1000. The level of schooling among the workers was not very high, but since reading at the workshops was institutionalized, the general instruction and information that they received contributed to enhance their intellectual and political awareness. The workers were organized into guilds that grouped together the main specialties in the factory: rollers, (dependientes), strippers, banding and packaging workers and selectors. Leaders were chosen by balloting and acclamation.

The H. Upmann workers took part in the 1902 strike for open access of Cubans to better wages, the 1907 strike for wages in US currency, the 1908 strike against cuts in the payrolls, the 1932 strike against cuts in wages and the 1949–1950 strikes against the introduction of mechanization in the industry that caused massive lay-offs.

At present, H. Upmann still produces and exports its famous vitolas plus other newly incorporated brands.

**Inside the present Partagas factory, these labels are used to adorn the wall.**

**(Above and below) Romeo y Julieta outside labels for the end of the box.**

# ROMEO Y JULIETA

In 1873, two Spaniards from Asturias, Inocencio Alvarez Rodriguez and Jose "Manin" Garcia Garcia, officially registered a brand of cigars named Romeo y Julieta, after the well known characters of Shakespeare's immortal tragedy. The factory where these cigars were made was located at No. 87 of the teeming San Rafael St. in down town Havana. It was a small factory but the efforts and dedication of its industrious owners soon propelled it to the forefront of the highly competitive market of export quality Havanas.

The original manufacturers of Romeo y Julieta prospered swiftly due to the high quality, superb finish and exquisite packaging of their products. Romeo y Julieta became a frequent choice of demanding consumers, won over by the excellence of these pure Vuelta Abajo cigars.

Eventually, Alvarez, Garcia and Co.—which was the commercial firm that manufactured the Romeo y Julieta, designed a successfully marketed new brands like La Mar, in June 1876, Los Amantes de Verona and Monteschi y Capuletti—to continue evoking Shakespeare's play—in June 1879, La Superfina, La Flor de Lozano Pendas y Cia., Daniel Webster and La Cubana, all in 1882, La Salamith, Entre las Rosas, La Mia, La Sonambula and Mary Stuart, in 1883. Ten years after its foundation the Romeo y Julieta factory had grown into one of the biggest and most popular producers of Havana cigars.

The secret behind such a remarkable success lay in the firm's meticulous selection of the tobacco leaves it purchased from among the finest harvests of the Vuelta Abajo region. For that pupose, the owners kept a number of highly skilled inspectors directly at the plantations who constantly sent back detailed on-site reports on the conditions prevailing at each plantation in the best districts. The factory thus had the necesary information to decide which leaves to buy. Furthermore, the Romeo y Julieta cigar rollers were chosen from among the best in Havana.

In 1886 "Manin" Garcia withdrew from the firm. After a brief one-year association with another industrialist named Montero, Inocencio Alvarez continued as sole proprietor until April 1902, when he sold the factory over to Prudencio Rabell, Gabriel Acosta and Jose Ferro, of the firm Rabell, Acosta and Co. This new entity managed the factory for a very short time. In June 1903 it was bought by Jose Rodriguez Fernandez, "Don Pepin" another

**Don Pepin, also known as Jose Rodriguez Fernandez, an Asturian.**

Asturian who had been the overseas sales manager of the Cabañas y Carvajal factory and was regarded by many in the trade as the last representative of the "Golden Age" of the Cuban cigar industry.

Don Pepin founded the firm Rodriguez, Arguelles and Co. Inc. whose three other business partners were Ramon Arguelles Busto, Antonio Rodes and Baldomero Fernandez, all Spaniards. The new administration transformed Romeo y Julieta into one of the leading manufacturers in Cuba. Their innovative commercial ideas based on active publicity, market studies, sales promotion and the appointment of skilful special sales agents in key countries made Romeo y Julieta a very competitive brand abroad. The main merit for these results was undoubtedly Don Pepin's, a man with a vast experience in the production and marketing of Havanas. Born in Asturias in 1866, he was sent to Cuba at the age of nine to learn all about the tobacco trade with an uncle. He succesively worked in every echelon of the cigar industry and steadily rose to be overseas sales manager of the important Cabanas y Carvajal. He quit this firm two years after it was bought over by an American company and decided to go into business privately.

A new Romeo y Julieta factory had been erected on the broad avenue of Belascoain, in midtown Havana, on the site previously occupied by the bull-fighting ring in colonial times. The new administration immediately undertook its complete refurbishing and expansion. Cuban cigar factories had been noted for their magnificent buildings since early times, which made a Spanish Minister of Overseas Dominions characterize them as "lavish constructions." Romeo y Julieta was indeed included among the most noteworthy.

Jose Rodriguez travelled frequently to Europe, Latin America and the United States. His personal prestige as a successful businessman was based on the fact that he never ran out of markets for the anual 20 million cigars that his 1200 workers including 750 hand rollers, produced.

In 1909 Don Pepin registered the trademark Romeo y Julieta at the International Union in Bern, Switzerland, under the number 7792. The better part of his output was exported to the United States where the different types of Romeo y Julieta cigars became first-class favorites. The British market came second to the American and then followed Canada, Australia and several South American countries throughout which the special sales agents displayed their dynamic abilities. Romeo y Julieta cigars were also highly prized in continental Europe; France, Spain, Portugal and Switzerland purchased substantial quantities of these vitolas.

In order to promote his sales in Italy, Don Pepin decided to set up a kiosk made with rare Cuban hardwoods and built by skilled creole crafstmen. The spot chosen to erect the kiosk was the very House of Montagu, in the city of Verona, where the legendary lovers'drama had unfolded; a most apropiate setting for a product that bore their names. Visitors were presented with samples of the famous hand-rolled Havanas.

Talented, generous, elegant, and charismatic, Don Pepin enjoyed a great popularity both in Cuba and abroad, always willing to give his invaluable assistance to fellow businessmen in the trade. His very special contribution to the international launching of the brands "El Crepusculo" and "La Gloria Cubana," of the firm J. Rocha and Co. was widely commented at the times.

The famous spanish poet Federico Garcia Lorca mentioned this outstanding cigar brand in his "Son de Santiago de Cuba" when he wrote: "Y en la Rosa a Romeo y Julieta Iré a Santiago.

**The well known Romeo y Julieta drawing on the label of the cigar box. (Enlarged 15%).**

In 1940, at the peak of his success, Don Pepin dissolved the firm Rodriguez, Arguelles and Co. and founded the Romeo y Julieta Cigar Factory Ltd, naming his nephew Hipolito Rodriguez as Vice Chairman. By then, his registered trade marks included Romeo y Julieta, Don Pepin, Falman, Flor de Rodriguez, Arguelles y Cia., His Majesty, La Mar, and Maria Guerrero, the latter created with the consent of the famous Spanish actress whom Don Pepin had met during one of his trips to Spain. Jose Rodriguez died a very wealthy man in 1954 at the age of 88, after a lifetime dedication to the cigar industry producing Havanas of the highest quality.

Throughout its history, the Romeo y Julieta factory created more than two thousand kinds of cigars of superior and exceptional quality and received the highest awards at numerous expositions:

Antwerp in 1885 and 1894, Brussels in 1888 and 1889, Melbourne in 1888, Paris in 1889 and 1900, and Liege in 1907. The famous "Churchill" vitolas were manufactured to cater to the tastes of Sir Winston Churchill, the British Prime Minister during World War II, who was an incorrigible smoker of Havanas. On one oration, commenting on his attachment to the island, Churchill retorted:

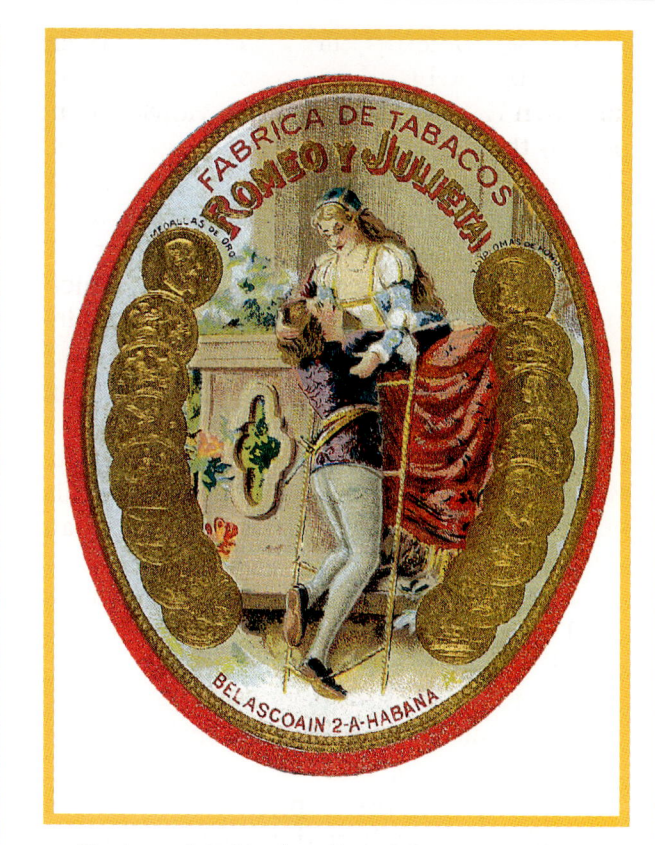

**Paste-on label for bundles of Romeo y Julieta.**

"I always carry Cuba on lips."

At present its universally acclaimed vitolas are manufactured and exported with great success.

**Note how this label differs from the one on the facing page? This label states: *Exclusively by Hand Elaborados a mano*. Does this indicate the box utilizing the label on the facing page contain machine made cigars? (Enlarged 10%).**

MADE IN HABANA, CUBA
EXCLUSIVELY BY HAND          ELABORADOS A MANO

117

GENERAL CHARLES DE GAULLE
LA CORONA
S. E. RENÉ PLEVEN
LA CORONA
LA CORONA
GS
HABANA
HABANA
HABANA
SPECIALLY SELECTED FOR
LA CORONA HABANA
A. BAYLIN, LONDON
LA CORONA
WR
HABANA
LA CORONA
WM F KENNY
HABANA
SPECIALLY MADE FOR
LA CORONA HABANA
THE MANHATTAN CLUB
SPECIALLY MADE FOR
LA CORONA HABANA
J. DUNCAN & SON
LA CORONA HABANA
COURT CORONA
LA CORONA HABANA
A. BAYLIN LONDON
SPECIALLY MADE FOR
LA CORONA HABANA
J. DUNCAN & SON
LA CORONA DE ALVAREZ Y LOPEZ HABANA
LA CORONA
CLUB DE LA UNION
HABANA

## LA CORONA

One hundred amd fifty years ago, in 1845, **Jose de Cabarga y Cia.**, owners of a cigar factory at No 129 Cuba St. in Havana, introduced the brand **La Corona** the "crown," which was to become one of the most famous cigar trademarks in the world.

In the years that followed, La Corona became increasingly popular among smokers due to the exquisite taste and fragrance of vitolas made with superb blends of Vuelta Abajo leaves. The brand soon won great fame in Cuba and abroad, where its cigars were in constant demand. In 1874 the factory was located at 95 Galiano St. in Havana.

Thirty-seven years later, in 1882, **Manuel Lopez y Cia.** another experienced manufacturer bought the **La Corona** brand from the widow of Cabarga and registered **La Corona** to his name together with the brands **Puck**, **La flor de Manuel Lopez**, **Alhambra** and **Jose Domingo**.

At that time Manuel Lopez Fernandez was the owner of **La Vencedora** cigar factory, which was located first in the Calzada (de Jesus) del Monte and in 1885, at No. 28 Figuras St. in Havana City.

In that same year of 1884, Manuel Lopez sold his factory to Segundo Alvarez, another outstanding personality in the trade since 1855, who then went into partnership with Perfecto Lopez, the owner of the "Katherine and Petruchio" cigar brand. Together they opened the firm **Alvarez, Lopez y Cia.** and henceforward exploited La Corona as well as the other brands. The new management moved the factory to the former **Palacio de Aldama** in Amistad St. between

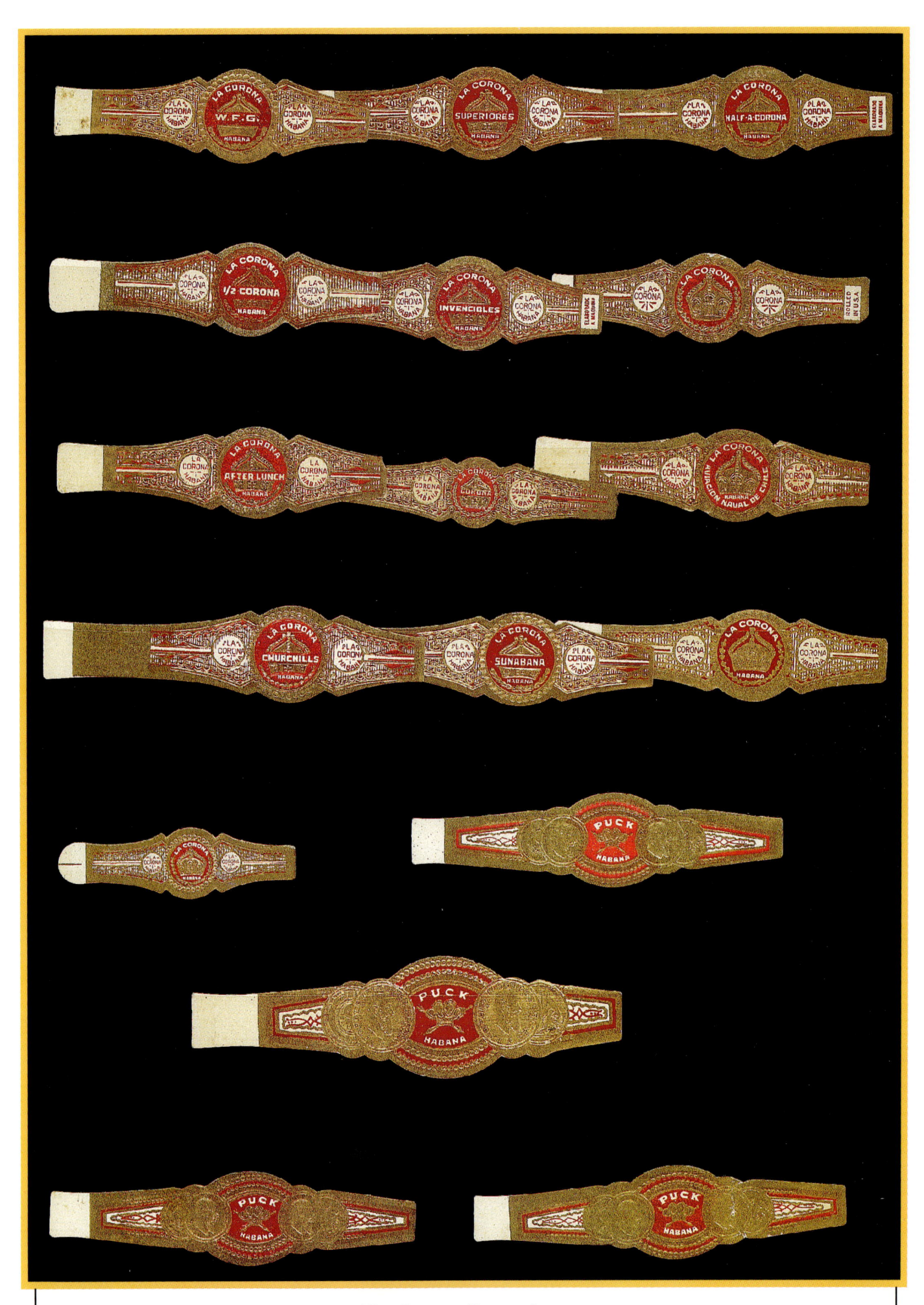

Miscellaneous Havana rings.

Miscellaneous Havana rings. (Enlarged 10%).

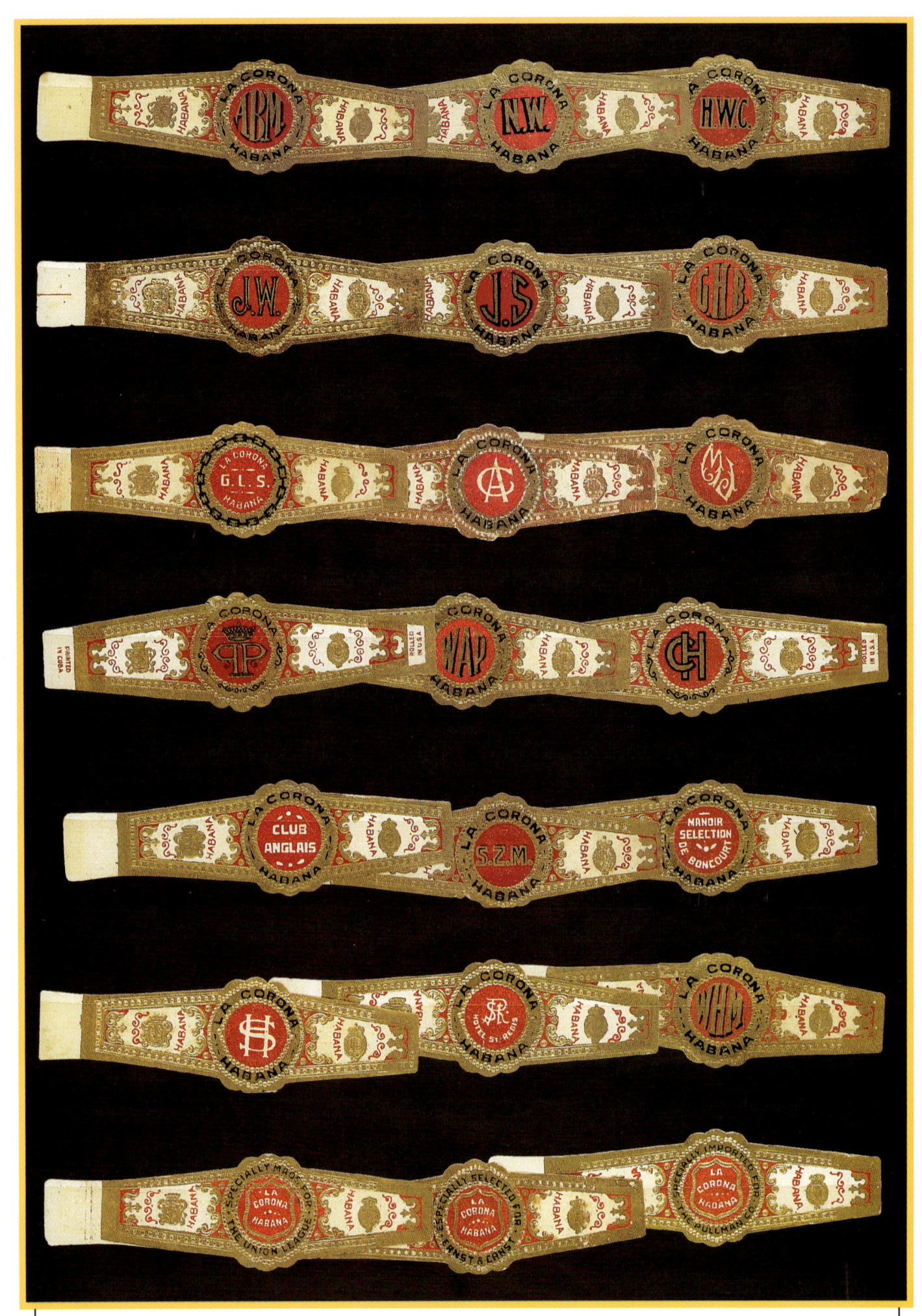

Miscellaneous Havana rings.

Miscellaneous Havana rings.

Miscellaneous La Corona vanity rings.

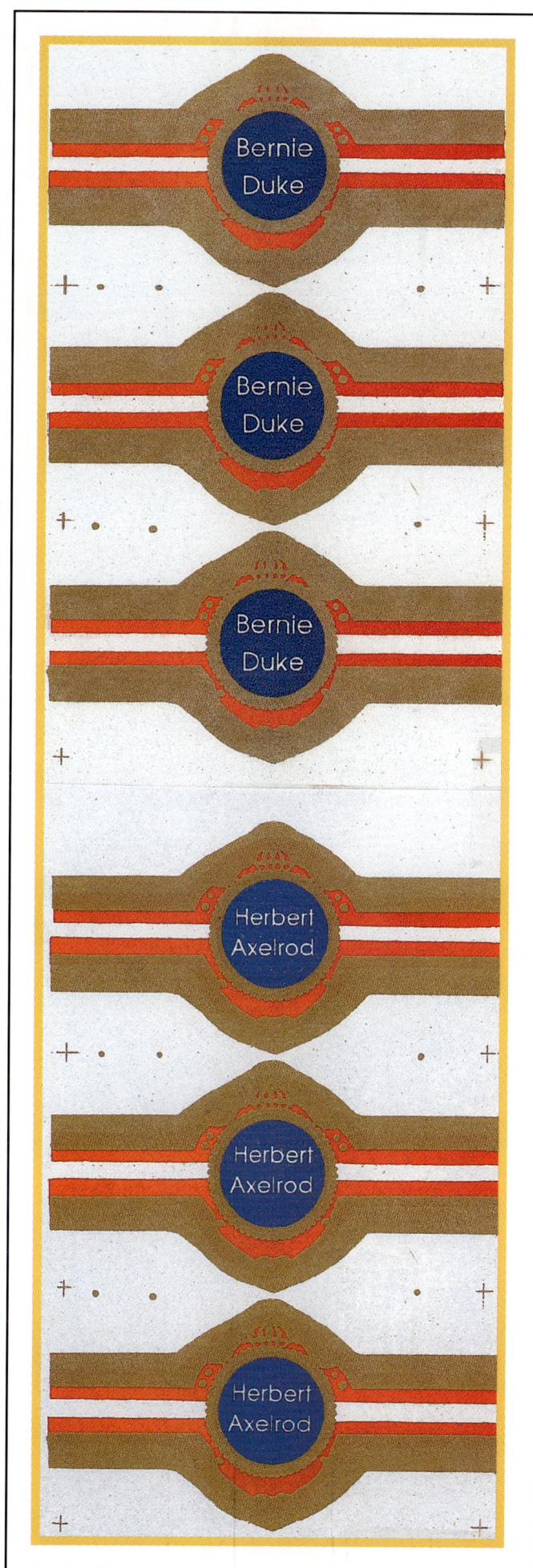

**Vanity rings made in Havana as a special gift in 1994. (Same size).**

Estrella and Calzada de la Reina. At the new premises and under Segundo Alvarez's dynamic and capable leadership and Jose Garcia's vast experience in the trade. La Corona reached universal recognition as one of the best and biggest manufacturers of Havana cigars achieving universal recognition.

In December of 1884 **Segundo Alvarez y Cia.** applied for a change of registration of **La Corona** ands in 1885 registered two new brands: **Las Coronas** and **Las Tres Coronas.**

Late in the eighties the brand's productive and commercial boom seemed endless, in spite of the fact that its new proprietor also shared the management of the Henry Clay factory that had been bought by English capital after the passing of its original owner, Julian Alvarez, and merged into the London based **Henry Clay & Bock and Company Ltd.** in 1888. But aside from managing both factories, Segundo Alvarez was the chairman of the **Union de Fabricantes de Tabacos de la Isla** the cigar manufacturers' union.

The vitolas by La Corona ranked among the best in the world for their quality, fine craftsmanship, and excellent taste and aroma. In 1898, Segundo Alvarez sold the factory with its 18 annexed brands to the English company **Havana Cigar and Tobacco Factories Ltd.** Shortly after that, George Washington Duke's **American Tobacco Company** bought La Corona and appointed Eustaquio Alonso as general manager.

In the year 1900, when the factory boasted a labor force of more than 600 rollers, 200 women workers and 200 apprentices, the firm registered three new brands: **La Flor de la Corona, Mi Corona and Dos Coronas** Ten years later, the factory's daily output of 40 thousand cigars was worth an estimated $800,000 per annum. The brand **Corona Imperial** was registered in 1910.

During the first 30 years of this century La Corona achieved its highest production rates ever: over ten million cigars a year.

1925 was the peak year with 39 million. A key role in this boom was played by Emilio Rivas, a cigar maker with more than 50 years in the trade, who was the factory's chief blender since the beginning of the century and had achieved one of the best Havanas ever produced. This brought La Corona the well-deserved fame as the international cigar "par excellence."

The world economic crisis of the late twenties and early thirties caused the consumer markets to shrink, especially in the United States, which was the export destination of the

Miscellaneous La Corona vanity rings.

La Corona vanity rings.

Vanity rings made by La Corona.

better part of La Corona's output. Of the 18,000,000 cigars manufactured in 1933, the firm managed to sell only 5,000,000.

Economic woes touched bottom that year and the Cuban cigar industry faced one of its most dramatic moments. The owners of La Corona came up with a plan to slash prices in order to boost sales, since Havanas were too expensive for the American consumer and a considerable number of smokers could not afford them. The management analyzed that the production of one thousand cigars in Cuba was costing them $54; the import and sales tax in the US was $127.50, plus $87 for the raw material, which brought the final cost to $268.90 per thousand cigars.

They finally came to the conclusion that costs could only be reduced in the manufacturing process by cutting wages and streamlining the worker's cigar allowance, the **fuma** which they figured was costing them $300,000 a year.

Faced with this option, the workers flatly rejected the wage cuts and went on strike. The American Tobacco Company then decided to move the manufacture of export quality cigars to the United States. The tobacco leaves would continue coming from Cuba, but women workers trained in the US and who were not interested in the **fuma** would be used as cigar rollers. These measures would certainly bring down the production costs and they would also mean added savings in taxes for the import of manufactured cigars.

The management had been planning ahead for this move. In Trenton, New Jersey, they had built a model factory with very similar facilities to La Corona's original ones in Havana. In 1933 they transferred manufacture of all the brands that were destined for export to Trenton. Their expert blender, Emilio Rivas, was also sent over to the United States. Meanwhile, the production of cigars for the internal market remained in Cuba in the hands of a newly formed company, **Tabacalera Cubana S.A.** which took charge of the domestic manufacture by virtue of Writ No. 184 of 21 June, 1932, signed in the presence of Virgilio Lasaga Castellanos, notary public, in the city of Havana.

Thus, the original premises that La Corona had occupied since 1882 in the **Palacio de Aldama** were dismantled and the factory moved to the building on the corner of Zulueta and Colon Streets, where the American Tobacco Co. had been concentrating its other cigar factories since the beginning of the century. All of these assets were passed over to the **Tabacalera Cubana S.A.**

La Corona soon took the lead over the other factories in the new premises, which was named after the world-wide famous brand. Its vast galleys were the scene of many workers' actions in demand for better living conditions, against the mechanization of the industry and the imposition of corrupt union leaders who had not been elected by the workers.

Today the building bears the name of Miguel Fernandez Roig, in memory of the La Corona workers' leader who was murdered there in 1948. It is still one of the main cigar factories in the country and continues to manufacture its famed Havanas.

**The La Corona Building.**

Edificio de "La Corona."

Vanity rings.

Vanity rings by La Corona.

131

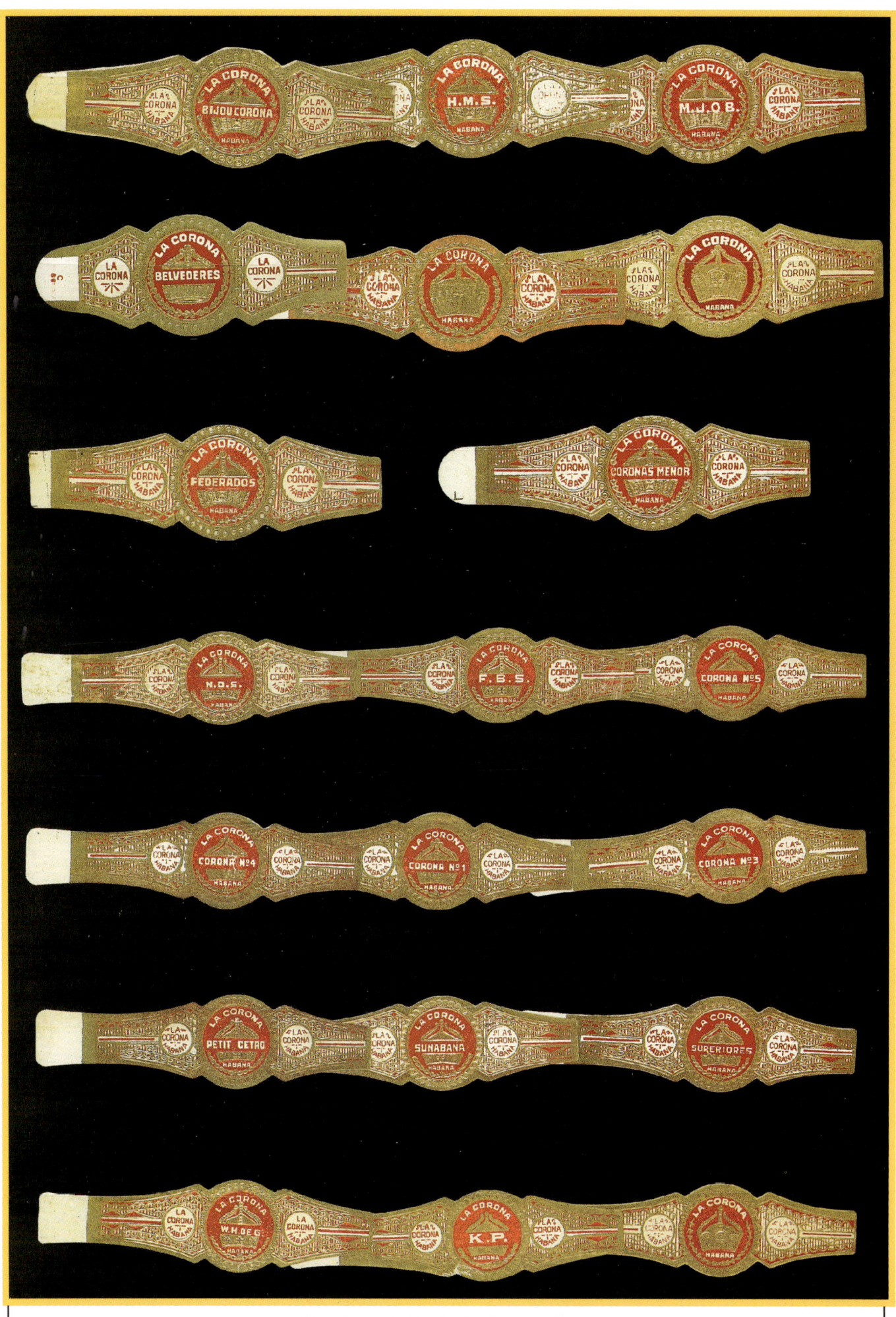

Cigar bands indicating La Corona sizes.

Special La Corona brands.

Various additional La Corona bands.

La Corona vanity bands.

**Francisco E. Fonseca adorns his label with spot illustrations of the Statue of Liberty in New York Harbor and the Morro Castle at the entrance to the Havana harbor.**

## FONSECA

The factory was founded in 1891 by Francisco E. Fonseca, the man who first marketed cigars in individual metal containers. In 1910 the factory was located at 128 Gervasio St. in Havana but four years later it moved to 102 Galiano st., where Manuel Valle used to have a cigar factory.

Fonesca was considered an artist in the cigar industry for his ability to create numerous deluxe vitolas. He had a solid reputation as a master in every aspect of his trade and possessed an exquisitely fine taste selecting tobacco. The quality of the leaves that went into his cigars was truly exceptional which allowed Fonseca to achieve a particular taste that has remained unaltered through the years.

Mr. Mecallin, widely reputed as the best cigar roller ever born, for many years headed the roller's department of this exclusive brand. The factory had a parlor or "studio" where the customers met to express their ideas, tastes and desires regarding the vitolas. Fonseca himself paid careful attention to this customers' suggestions and almost invariably satisfied their wishes.

Each Fonseca cigar was packed in a metal tube and wrapped with find Japanese paper. This double insulation protected the cigars against damages, changes in the weather, excessive dryness or humidity, dust, scent and flavor contamination, and form any direct contact with the hands of salesman.

More than 60 % of the cigars produced by Fonseca were made for receptions, banquets and conventions or as presents and souvenirs. The rest generally went to exclusive clubs and discerning smokers fond of all things noble.

The growing popularity of this brand was confirmed by the number of persons from all over the world who expressed their satisfaction to Francis E. Fonseca for the quality of his cigars and for the artistically refined presentation of his products.

Fonseca remained as an independent producer and in 1909 he registered the band **Hamlet** and in 1914 **"La Flor de Fonseca."** After the economic crisis of the Twenties and the crash of 1920, he joined in partnership with Tiburcio Castaneda and J. Montero and founded **Castaneda, Montero, Fonseca S.A.**, located at 466 Galiano St. This firm appeared registered in the **Registro de Fabaricantes Exportadores** of the **Comision Nacional de Propaganda y Defensa del Tabaco Habano** in 1940, 1951 and 1954 as owners of the brands **Castaneda**, **El Genio**, **Filotea**, **Fonseca**, **Hamlet**, **J. Montero y Cia**, **Lurline**, **Para Mi**, **Real Carmen** and **Rotario**. In his celebrated poem **"Son de Santiago de Cuba"**, written in Havana in 1930, Federico Garcia Lorca consecreated this brand of cigars when he wrote:

*"Con la rubia cabeza de Fonseca*
 *ire a Santiago*
 *Mar de papel y plata de moneda*
 *ire a Santiago "*(*)

He alluded thus to the embossed points that embellished the cigar boxes, sent directly from Havana, which he had seem since childhood in his father's house. Keenly recorded in his mcmory, those prints on the cigar boxes gave the future poet a vivid idea of Cuba.

At present, "Fonseca" cigars are produced with the same quality standards and go out mainly to the Spanish market, to the complete satisfaction of countless clients.

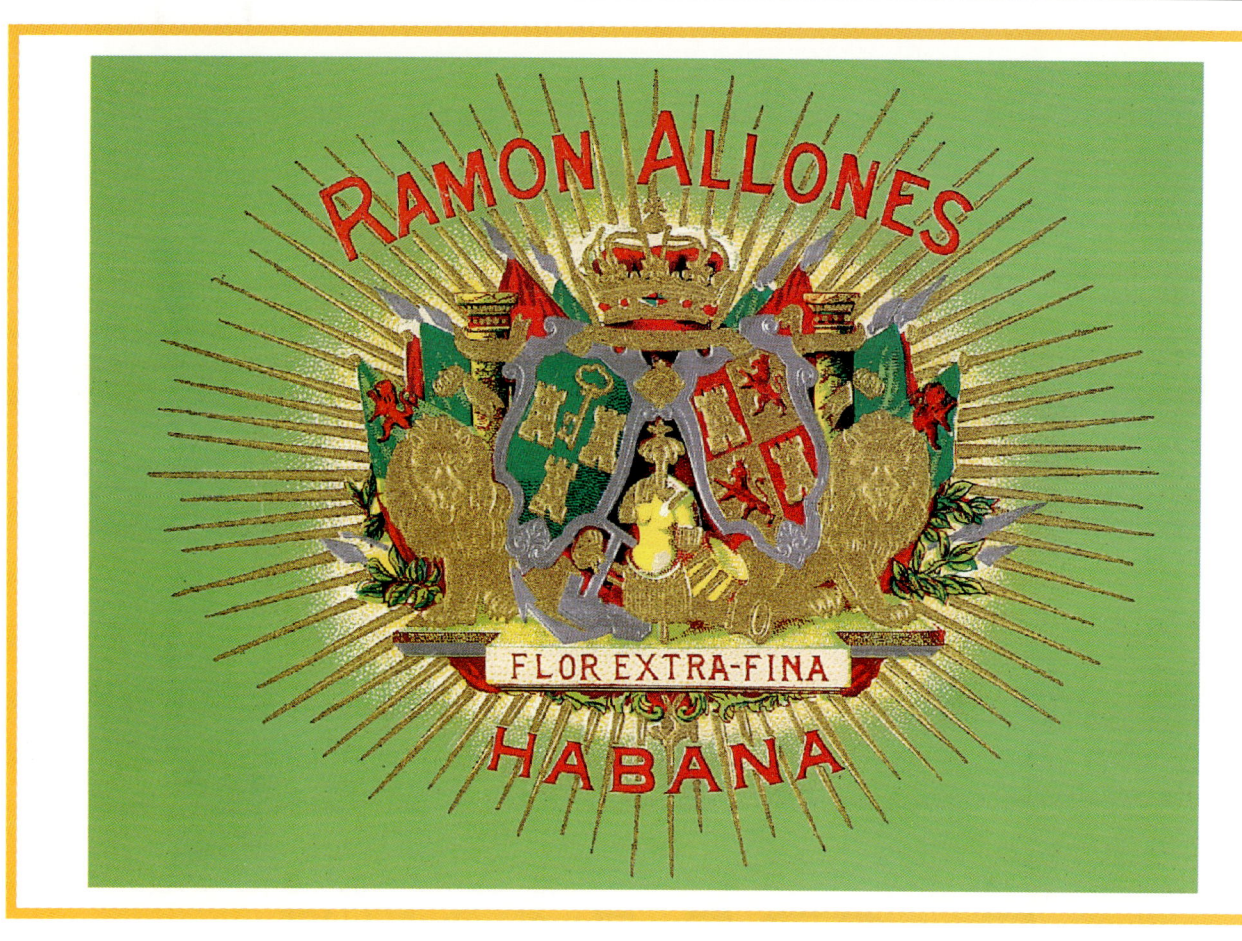

Ramon Allones box label.

## RAMON ALLONES

This famous brand of Havanas was established by the Galician Don Ramon Allones and his brother Antonio in 1846, at his factory **La Eminencia**, located at No. 129 Animas St., Havana.

In his endeavors to enhance the quality and fine packaging of his cigars, Allones was the first manufacturer to introduce the superior vitolas or **regalias** and in the late fifties of the 19th century, the luxury cases that won so much praise at the tables of emperors, kings and princes alike. It is said that the sumptuous cases and boxes developed by Allones were but the appropriate containers for the exclusive Havana cigars.

A true expert in matters of tobacco, Allones achieved outstanding blends of Vuelta Abajo leaves to produce exquisite **regalias** that were soon coveted by smokers of **puros**.

Among the demanding clients of the brand was the Royal House of Spain, of which Allones became an appointed supplier with the right to use the coat of arms of the kingdom in his sophisticated **habilitaciones**.

By 1866, Allones had also acquired the brand **El Designio** while the demand for his cigars in the European, North American and South American markets experienced a sustained boom.

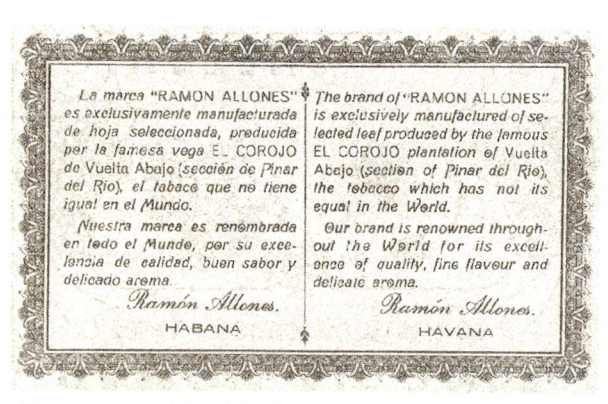

The reverse side of the box label bears this printing.

Bundle label for Allones.

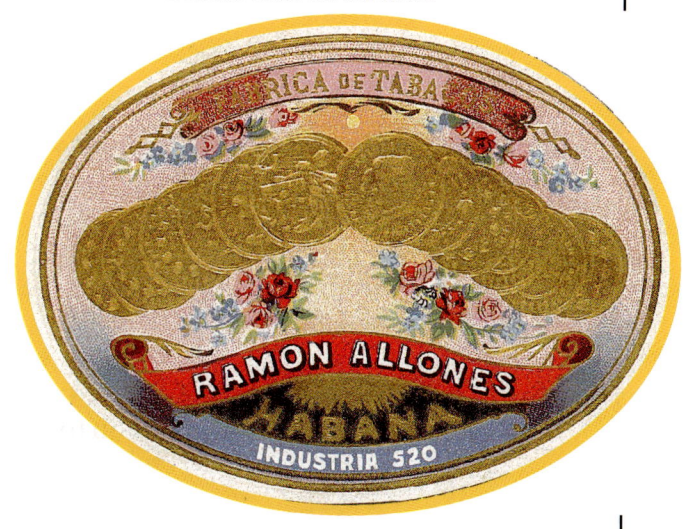

137

**A poorly printed box label for Ramon Allones. The gold is faked and not embossed.**

Many years later, in 1910, the **La Eminencia** factory had gone over to **Rabell, Costa, Vales y Cia.**, located at No. 98 Galiano St. The following year British investors bought the firm and continued to operate under the name **Allones Limited**. Alonso Allones and Florentino Romero Allones, both nephews of the founder Ramon, were kept on as managers at the new firm.

In 1927, **Ramon Cifuentes y Cia**, the owners of **Partagas**, bought the brand and incorporated it into their factory. Since then, the **Ramon Allones** vitolas were manufactured under the expert supervision of these fine cigar makers who zealously preserved the particular traits of the famous brand.

**Allones ring.**

Among the most distinguished vitolas by **Ramon Allones** are **Seleccion Especial, Coronas, Coronas Gigantes, Petit Coronas, Small Club Coronas, Panetelas, Churchills,** and **Ramonitas**, which are still manufactured at the Partagas factory.

*Printer's proof of Sol de Cuba box label. (Same size).*

## SOL

The Sol factory was established in 1840 at No. 91-93 Consulado St. in Havana by its founder, Marcelino Borges, who used only the best Vuelta Abajo leaves in his blends.

Borges exploited this brand until his death in 1860. A period of decadence followed until it was bought by **Behrens y Cia** in the final years of the eighties. Behrens and his two partners, who had considerable experience with other cigar factories, revitalized production and in a few years they raised Sol cigars back to first-class brands and to outstanding public demand. Sol was awarded a gold medal at the Antwerp Fair in 1894.

Sol was one of the so-called independent factories and by 1910 with a labor force of 400, it was producing some 35 thousand cigars a day, while its annual output was close to 10 million. The brand had special representative in New York, Mr. Max Schatz, whose office was located at 76 Pine Street.

The main brand of the Sol factory was **Luis Marx**, named in 1883 after the famous tobacconist owner of numerous plantations in Alquizar, province of Havana, and who later became a buyer for the American Trust.

It is said that Sol was the first brand of cigars to be smoked by Commander Robert Peary when he arrived at the North Pole in 1909.

In 1940 the brand was owned by **Martinez y Cia**. and was located at No. 200 Real St. in the municipality of Marianao, in Havana. The factory continued producing its export quality vitolas there until 1960. Today the vitolas by Sol still rank high in the international market of Havana cigars.

*Luis Marx ring. (Enlarged 10%).*

*An elaborate Sol ring. (Same size).*

Sol bands in giant size. (Same size).

Sol bands. (Same size).

Giant Sol bands.

141

Sol box label. (Enlarged 80%).

Sol rings.

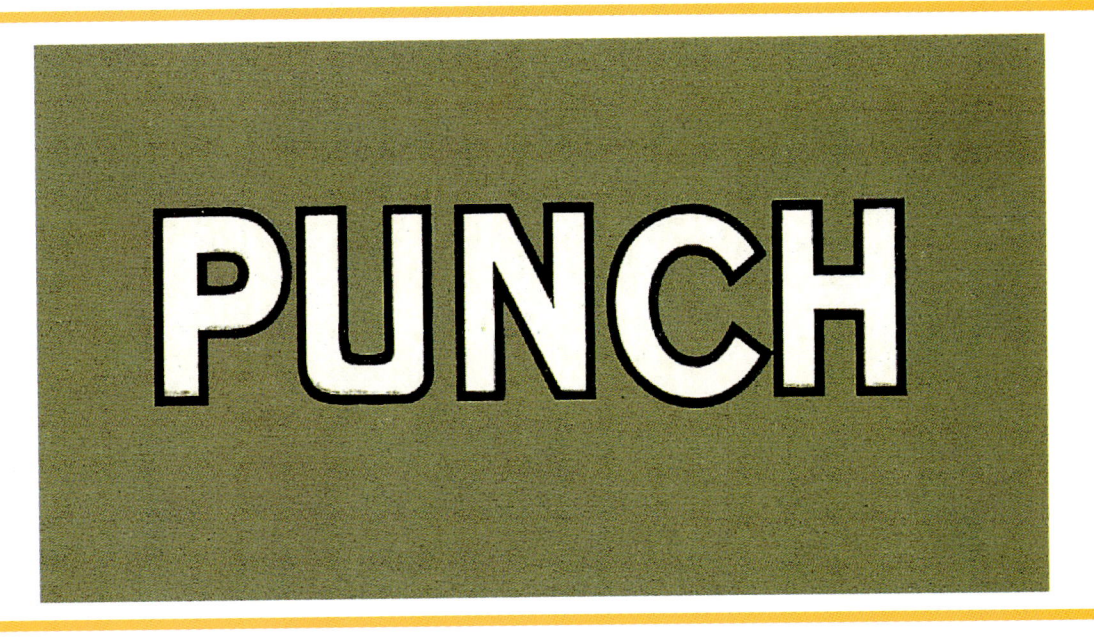

Punch box label.

## PUNCH

This famous brand was created by **J. Valle Cia** in 1840, at a time when other legendary makes of Havana cigars were on the rise. Since the brand was originally conceived for the British market, Juan Valle named his cigars after the popular satirical magazine.

In 1874 ownership of **"Punch"** has gone to Luis Corujo, who owned a cigar factory at 38 Gervasio St, and the brands **"El Comerciante"**, **Camarioca Flor de Corujo, Flor de J.C. Ruiz, Hija del Regimiento** and **"La Sin Par."**

In 1885 it was registered by **Lopez y Fresqueres**, owners of the brand **La Camarioca**. **"Punch"** began to appear among the distinguished brands of Havana cigars in the guide that the **"Grand Hotel Pasaje"** published for his guests. As of 1902, Manuel Lopez Fernandez remained as its sole proprietor and became one of the independent cigar manufacturers. The factory was located at No. 28 Rayo St. in Havana.

Manuel Lopez was a wealthy and industrious Spaniard with long years in the tobacco trade who, since 1880, owned the **Las Vencedora** factory located on Monte Street. Lopez revitalized **Punch** and achieved great success in the European markets, mainly in Great Britain and Spain where its vitolas came to be highly regarded for their mild taste.

In 1924 after Manuel Lopez decided to retire from the tobacco business Esperanza Valle Comas became the new owner of Punch. Years later she sold it to **Fernandez Palicio y Cia Inc**. who since 1940 registered Punch as one of their 19 export brands and one of the 63 that Spain used to import at that time.

Among the numerous vitolas by **Punch**, made with premium blends of the best Vuelta Abajo leaves, some of the best known are **Punch Punch, Petit Punch, Panetelas, Corona, Doble Corona, Gran Corona, Petit Corona, Diadema Extra, Royal Selection No. 11 and 12, Seleccion de Luxe No. 2, Churchill, Super Seleccion No 1 and 2,** and **Margarita, Amorosos, Nacionales, Ideales de Punch, Gloriosos, Ninfas, Brevas a la Converva, Triunfos, Cremas Superfinas, Capitolios, Rayados, Exquisitos, Perfectos, Principes, Punch Cocktails, Rosemarys, Belvederes, Petit Presidentes, Petit Elegantes y Privcisas Finas.**

At present, the vitolas by Punch are as fine as ever and in the traditional markets their countless fans can find them.

Punch box label. (Same size).

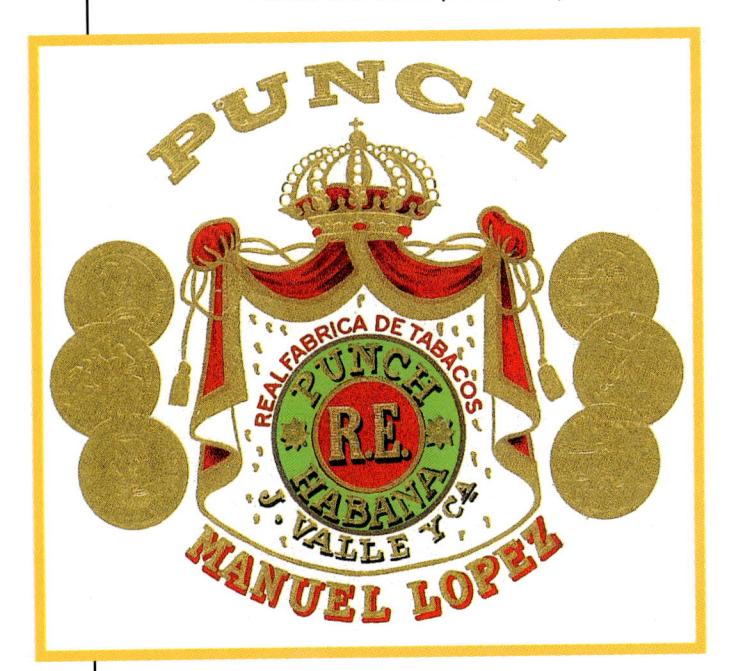

# PUNCH
## GRAN FABRICA DE TABACOS
### DE J. VALLE Y Cª, HABANA
### MANUEL LOPEZ

## MADE IN HAVANA, CUBA

Punch box label. (Same size).

Lonsdale of Manuel Lopez. (Enlarged 45%).

Bolivar box label.

# BOLIVAR

Although some researchers believe that the brand **Bolivar** had an early start in this century, the first official record of its existence is dated 1921, when **J.F. Rocha y Cia**. applied for the registration of a brand of cigars with the name **Bolivar**. The application was approved and the firm received the corresponding Certificate No. 38374 of the Trademark and Patent Office of the Ministry of Agriculture, Commerce and Labor of the Republic of Cuba.

**Bolivar** was owned in equal shares by Jose Fernandez Rocha, jose Rodriguez Fernandez and Robert Middenas and its registration was successively renewed in 1923, 1933, 1940, 1951 and 1954 by **J.F. Rocha y Cia**. Its first location was No. 100 San Miguel St. until it moved to No. 364 of that same street in Havana. In 1944 the brand "La Mercantil" was incorporated to "**Bolivar**".

The name **Bolivar** is a tribute to the liberator of Venezuela, Columbia, Peru, Bolivia and Ecuador from Spain's colonial rule in the first three decades of the 19th century. His full color effigy appears in the superb "habilitaciones" (label on the box) that decorate the boxes and bands of each Bolivar cigar.

The vitolas of Bolivar are noted for their dark wrapper and robust aroma. Among the best known are **Coronas, Royal Coronas, Coronas Extra, Coronas Gigantes, Petit Coronas, Coronas Junior, Belicosos Finos, Palmas, Inmensos, Bonitos**, and **Gold Medal**. Bolivar cigars are now manufactured at the Partagas factory in Havana and are exported regularly with the same distinguishing quality that made them famous.

**BOLIVAR**

**BOLIVAR**
25 - PETIT CORONAS   Made by Hand

**BOLIVAR**
25 - PANETELAS   Cellophane

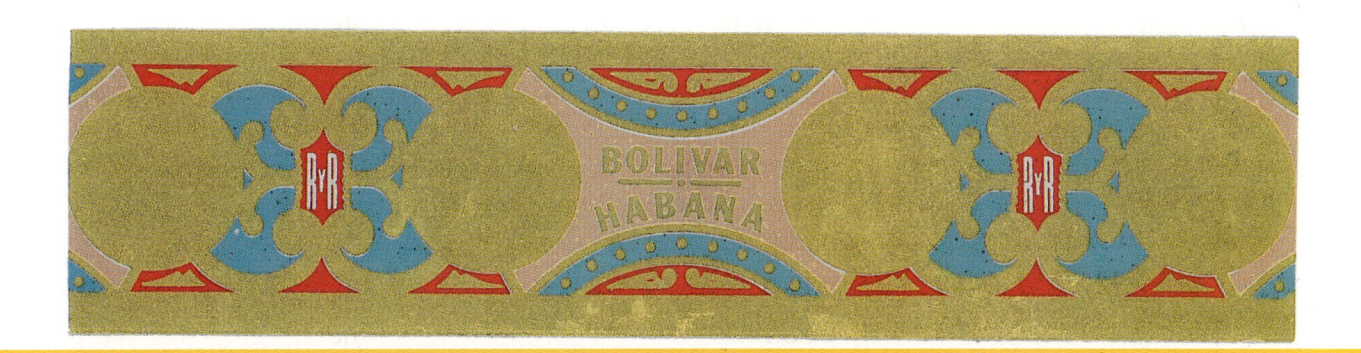

BOLIVAR HABANA

Box end labels and decoration. (Enlarged 50%).

# FABRICA DE TABACOS
# SAINT LUIS REY

San Luis Rey box label. (Enlarged 50%).

## SAINT LUIS REY

This brand was created in the late thirties by **Zamora y Guerra** at the request of British tobacco importers Michael Keyser and Nathan Silverstone. In 1940 the brand was already listed at the Registry of Exporting Manufacturers of the **Comision Nacional de Propaganda y Defensa del Tabaco Habano** and was located at No. 810 Maximo Gomez St. in Havana.

Exquisite taste and a highly refined aroma characterize the cigars by Saint Luis Rey. Its vitolas are made with a selection of Vuelta Abajo leaves using wrappers which are normally dark in color and smooth in texture. Its **Coronas, Petit Coronas, Lonsdales, Regios, Serie A**, and **Churchills** were in great demand in the British and American markets. There is also a **"Saint Luis Rey"** brand with different markings that is manufactured in Cuba specifically for the German market.

Today the Saint Luis Rey vitolas are manufactured at the former Romeo y Julieta factory and are still exported to their traditional markets: Great Britain and Germany.

Jose Suarez Murias box label.

## LA DEVESA DE MURIAS

This famous brand of Havana cigars was founded in 1882 by Felix Murias Rodriguez, member of a family of tobacco businessmen. Years later, his brother Pedro, also an experienced tobacconist, took charge of the factory and managed it until his death in 1906.

Pedro Murias had established an important cigar factory in the corner of Zulueta and Apodaca Streets, in Havana. His cigars won early fame for their excellence and variety, which opened them the doors to the main European markets. In the nineties, Murias was already selling close to six million of his numerous brands of cigars in the Old Continent including special supplies to the Royal House of Spain.

In those years, Pedro Murias registered more than 20 different brands of hand rolled cigars; among those were **El Palacio de Cristal, La Prima de Murias, La reserva, La Warda, La Soltera, Walter Scott, La Opulencia, La Indiferencia, American Flag, La Paz de Cuba, Flor de los Campos de Cuba, La Joven America, El Idolico, Balmoral, La Prieta, La Manteiga, La Meridiana, Pedro Murias y Cia.,** and **La Flor de Murias**, almost all inscribed between 1896 and 1897. His operational capital was estimated at $1,000,000 at the time.

Many of these brands were bought by the Henry Clay & Bock and Company Ltd. and by the Havana Commercial Company between 1899 and the initial years of this century.

However, after the passing of Pedro Murias, **La Devesa** went over to Eduardo Suarez Murias who then leased it to Manuel Campos Rivero, a famous and wealthy pioneer of the Havana cigar industry. In 1904, Campos Rivero was a supplier to the Royal House of Spain and the owner, together with his partner Casimiro de lost Prados Nicolas, of the brands **Elsa, La Extrafina, La Flor de Donato Campos, La**

**Diosa de Cuba, La Nina Cubana**, and **Hermosa**. Campos was also the proprietor of the **El Retiro** plantation in the highly exclusive tobacco district of San Luis, in Vuelta Abajo, Pinar del Rio, which provided him with excellent tobacco to guarantee the sustained quality of his vitolas and brands.

In 1910, Campos Rivero was manufacturing **La Devesa de Murias together with the brands La Viajera, La Suiza Espanola, La Aurora**, and **Baire**. His vitolas were highly regarded in Spain, England, Germany and Portugal where special representatives of the firm were posted. Other big importers were the United States, Russia, Argentina and Chile.

In 1919 Eduardo Suarez Murias again recovered the brand and leased it to the Havana Commercial Company. A few years later, this enterprise transferred the manufacture of **La Devesa de Murias** cigars to the territory of the United States, but the markings on the boxes continued to bear the label "Made in Havana, Cuba."

In 1928, the **Comision Nacional de Propaganda y Defensa del Tabaco Habano** notified the leasing company of the prohibition against any form of markings indicating a false origin on cigars that had not been manufactured in Havana. The **Comision** also arranged with Eduardo Suarez Murias, the owner of the brand, to cancel the leasing contract with the Havana Commercial Company on grounds that the brand was registered in Cuba to identify cigars that were rolled by natural persons or entities listed as residents in the country.

Suarez Murias agreed to the request and cancelled his contract with the Havana Commercial Company. Eventually, he leased the brand to the firm **Martinez y Cia.**, located at No. 200 Real St. in Marianao, that kept it among its 19 registered trademarks until 1960.

Murias label. (Enlarged 80%).

La Gloria Cubana box label.

La Gloria Cubana ring. (Same size).

La Gloria Cubana. Bundle label.

# LA GLORIA CUBANA

Jose Fernandez Rocha and Jose Rodriguez Fernandez, who operated the firm **J.F. Rocha y Cia. Inc.** located at No. 36 San Miguel St. in Havana, registered **La Gloria Cubana** in 1920. Made with carefully selected leaves from the best plantations in Vuelta Abajo, this brand of cigars achieved immediate success.

Since 1887, **J.F. Rocha y Cia. Inc.** was the owner of **El Crepusculo** which had been founded by Manuel Castro Fuentes and Cesareo Garcia Cabanas in 1882.

The 1940 Registry of Exporting Manufacturers of the **Comision Nacional de Propaganda y Defensa del Tabaco Habano** showed that eight brands had been registered by this firm: **Bolivar, Flor de Ambrosia, El Crepusculo, La Gloria Cubana, La Navarra, La Petenera, Nene,** and **La Glorieta Cubana**. By then Jose Fernandez Rocha had already passed away but the firm kept its name under the general management of Rene Rocha.

La Gloria Cubana was one of the 63 brands of Havana cigars that were regularly imported by Spain in the forties. Vitolas like **Cetros, Medaille d'Or 1, 2, 3, and 4, Minutos, Tainos, Tapados**, and **Sabrosos** had a particularly great demand.

For a number of years after its intervention in 1960, the factory's output dwindled. The manufacture of these strong and aromatic cigars is done today at the Partagas factory, especially the vitolas **Medaille d' Or.**, which are marketed in fine 8-9-8 cases.

## SANCHO PANZA

**Salvador Pareto y Cia**, located at 142 Manrique St. and owners of **"Jenny Lind"** and 19 other cigar brands, had **"Sancho Panza"** to their name in 1874.

**Allones Limited** registered this brand in 1920, but when seven years later **Cifuentes, Pego y Cia** bought or leased almost every brand owned by **Allones Limited, Sancho Panza** went over together with **Flor de Alma, Ramon Allones, Amor en Sueno, Guayarre, Algo Bueno, Mi Necha** and **Modelo de Cuba**.

In 1931, **Sancho Panza** was bought by **Rey del Mundo Cigar Co.** located at No. 852 Padre Varela St. in Havana, which held it until 1960.

The best known and most consistently demanded vitolas by **Sancho Panza are Bachilleres, Coronas, Coronas Gigantes, Non Plus Petit Coronas, Sanchos, Molinos** and **Panetelas Largas**.

**Sancho Panza** cigars are still manufacture for the foreign market and can be found especially in Spain and in France.

Sancho Panza bundle label. (Same size).

La Flor de Cano.

## FLOR DE CANO

The brothers Juan and Tomas Cano established their cigar factory in 1884. In 1891 they created the brand **La Flor de Cano** and manufactured fine cigars under that name in their small shop located at No. 615 Manrique St. in Havana.

Their vitolas possessed the excellent taste and exquisite aroma that characterized Havanas made with genuine Vuelta Abajo leaves. Their **Diademas, Coronas, Gran Coronas**, and **Short Churchills** were in great demand.

Cano was always an independent manufacturer. In 1940 the firm was registered as **Juan Cano Sainz** at the Registry of Exporting Manufacturers with the brands **Caracol, La Rica Hoja**, and **Trocadero**. In 1951 the firm changed its name to **Juan Cano e Hijo** and continued with the same brands in its factory at No. 301 San Gabriel St. at the corner of Durege, in La Vibora, Havana.

Vitolas by **La Flor de Cano** are still manufactured and exported mainly to the Middle East.

J.Cano poorly printed label. (Reduced to 85% of original size).

Bundle label for the Don Quijote de la Mancha label.

## DON QUIJOTE DE LA MANCHA

Juan Cueto was an expert in every aspect of the tobacco trade who in 1874 had a small factory at No. 31 Maloja St. in Havana. In partnership with J. Pedro he registered the brands **"La Flor de Navas"**, **"Flor de Canela"**, **"Flor de J. Venancio"**, **"Lord Rivers"**, **"Matilde"**, **"Mayerbeer"**, and **"Pelicano"**. In February, 1882 he registered two new brands **"Obeso y Cueto"** and **"El Dios Marte"**, under the firm **Juan Cueto y Cia.** In September that same year came **"Don Quijote de la Mancha"** and then **"El Gladiador"**.

In very little time, his cigars, made exclusively with blends of Vuelta Abajo leaves, achieved great success and Juan Cueto became a supplier to the Royal House of Spain. The latter was only one of the numerous European markets that imported his Havanas during the last twenty years of the 19th century and the first decade of the present century.

Eventually, in 1900 Cueto sold his factory to the Havana Cigar and Tobacco Factories Ltd. which in 1919 renewed the registration of the brand **Don Quijote** to its name, but sold it to the Tabacalera Cubana S.A. in 1932. The following year the Tabacalera again registered the brand which continued listed in 1940 and 1951 among the 91 registered brands of that enterprise at the Register of Exporting Manufacturers of the **Comision Nacional de Propaganda y Defensa del Tabaco Habano**, under Class 14 of the official nomenclature.

Box label for Flor de Juan Lopez.

## FLOR DE JUAN LOPEZ

In the 1870s Juan Lopez Diaz established an important cigar factory at No. 170 Industria St. in Havana, where he manufactured first-class Havanas with Vuelta Abajo leaves. His clientele grew steadily on account of the outstanding qualify of his vitolas which were already sought after in the international markets.

Following a time-honored tradition by every leading manufacturer, Lopez launched his most select product in 1876 and named it **La Flor de Juan Lopez**. In 1890 the brand had become so famous that it was listed as one of the top ranking Havana cigars in the souvenir guidebook that the **Gran Hotel Pasaje** gave its guests.

A skilled craftsman, Lopez created numerous brands of cigars which came to be highly regarded by the connoisseurs. In 1884 he registered **La Veneciana, La Mandolineta, El Marques de Caxias** and **Transcontinental**. The following year he registered **La Tarde, Facon, La Ritica, La Betica,** and **El Bello Aroma**.

Juan Lopez Diaz died in the initial years of this century. In 1910 his successors were still producing the vitolas of this brand at No. 6-8 Dragones St., Havana, where the brand **La Regia Inglesa** was also manufactured. Later on, La Flor de Juan Lopez was bought by a firm of independent manufacturers, **Cosme del Peso y Cia.** owned and managed by Cosme del Peso Perez and Abelardo Gonzalez Herrera, whose factory in 1924 was located at No. 314 San Ignacio St. in Havana, where they already manufactured the brands **Flor de Tomas Gutierrez, "La Flor de Diaz y Garcia", La Igualdad** and **Pierrot**.

In 1940, **La Flor de Juan Lopez** was still registered by **Cosme del Peso y Cia.**, together with the other brands, in the Registry of the **Comision Nacional de Propaganda y Defense del Tabaco Habano**.

The best known vitolas by **La Flor de Juan Lopez** are **Coronas, Petit Coronas, Patricios** which are nowadays manufactured and exported mainly to the European markets.

## BELINDA

In 1882, a new cigar factory began to operate at No. 96 and 98 Gervasio St. in Havana. On the 16 of June that year, the owner, Francisco Menéndez Martínez, an Asturian with ample knowledge of the tobacco trade, had registered at the Office of Industry and Commerce of the Higher Civil Government of the Island of Cuba, the brands **Belinda** (after which the factory was named) **Le Diamantina, La Flor de J. Alvarez** and **Hermandad**.

Those were the boom years of the Cuban cigar industry. Menéndez Martínez, known to his friends and relatives as "Don Panchin", began to produce the **Belinda** vitolas with excellent Vuelta Abajo leaves that he himself selected.

The market soon accepted Don Panchin's products and the **Belinda** Havanas began to be exported to traditional cigar-consuming countries like England, the United States, Spain, France and Germany. His factory became so famous that in 1885, Don Panchin registered a new brand: **Ramillete de Julio**. In 1890, the factory was listed as one of the most outstanding in Havana in the Guidebook that the Gran Hotel Pasaje gave out the their guests as general information about the city.

The Second War of Independence of Cuba (1895-1989), considerably affected the tobacco harvests, especially in Vuelta Abajo, causing a sharp decline in the availability of really good leaves. Like many other factories, **Belinda**'s output and financial resources dwindled. But when the harvests were again stabilized in the early years of this century, Don Panchin manages to pull through and recover the quality and standing of his cigars.

In 1980, as purveyor to the Royal House of Spain, and with the backing of **López y Cia.**, he registered his brand **Flor Fina Belinda**. The registrations was done by the firm "Fernández y Menéndez", because Don Panchin had been obliged to take in a new partner in search of much needed fresh capital for his factory.

In 1919 the firm was registered anew as **La Belinda, S.A.**, without any mention of Francisco Menéndez. The new brands **Royal Spirit** and **Adela** appeared. The following year, Manuel López Fernández, the owner of Punch, bought the new firm with all its brands and operated these until 1924. That year he sold them to Ramón Fernández Alvarez and Fernando Palicio Arguelles, of "Fernández, Palicio y Cia. Inc.", located at No. 51 Máximo Gómez St. in Havana, owners of the brands **Gener, La Escepción, Hoyo de Monterrey, Gioconda** and others. That same year, the new proprietors leased **La Belinda** to "A.R. Fernández y Hno.", owners of the brand **Hermani**, who ran **Belinda** until 1926.

"Fernández, Palico y Cia. Inc." again leased the brand to "Santiago y Hermano" until 1930, when they finally took direct charge of the factory and managed it uninterruptedly until 1960.

## QUINTERO

In 1924, Agustin Quintero and his brother founded a small cigar factory at No. 16 Delouet St. in Cienfuegos, in the former central province of Santa Clara, where they began to produce first class Havanas with leaves that came exclusively form Vuelta Abajo.

The brand was registered as **Quintero Hno.** and their mild, pleasant tasting cigars soon won a solid reputation among domestic and foreign smokers.

The remarkable stability of the blends that Agustin Quintero made contributed greatly to the brand's fame, which broadened his markets and boosted his business.

Spain was one of the big buyers of vitolas by **Quintero Hno.** in Europe. Way before 1940, it was one of the 63 cigar brands that Spain used to import from Cuba.

The firm **Agustin Quintero y Cia.** was listed in the Register of Manufacturers and Exporters of the **Comisión Nacional de Propaganda y Defensa Tabaco Habano.** The factory remained at 16 Delouet St. in Cienfuegos until 1960 where it produced the brands **Quintero Hno., La Riqueza** and **El Cañón Rayado**.

Today Cienfuegos is the capital of the province and the exquisitely mild **Quintero Hno.** cigars are still produced at the original factory.

## Chapter Fourteen
# HOW A HAVANA CIGAR IS MADE

Decades of industriousness were needed to develop a standardized method for making top-quality cigars as we know them today. That long-pursued method was the outcome of a gradual evolution from the somewhat clumsy procedures of the early times to the sophisticated and highly elaborate process that became the classic standard in all factories ever since the mid-nineteenth century.

When full-fledged factories didn't even exist and cigars were made either in private homes or in small workshops, tobacco leaves were simply rolled together and wrapped with another identical leaf.

This feature set them apart from the ones that were wrapped with corn leaves or paper and consequently were known as *pure tobacco* or simply *puros*, the Spanish word for pure. The drawing end of the cigar was twisted tightly without any glue forming a braid normally called *pigtail* which the smokers had to cut away or bite off. The resulting cigars were rather primitive and far from uniform; some were thicker and heavier than others and the color of their wrapper leaves—within a single bunch or package—were clearly different.

Nevertheless, the *puros* made in Havana were dearly appreciated by all those who tried them. The demand for *puros* or *Habanos* grew by the day and the commercial trade became a lucrative and booming business. Sailors and ship passengers who stopped at Havana's docks evolved steadily into a sizeable and cigar wise clientele.

Since 1818, when the free growing, manufacture and trade of tobacco was legalized, an extraordinary cigar frenzy occured. In Havana hundreds of people began to make cigars in their homes either individually or as a family job. The most daring hired several persons to manufacture cigars in makeshift workshops for a salary. These small independent producers generally worked in association with or financed by larger merchants who supplied the necessary means to produce cigars and then bought the finished products. In the nascent cigar-making industry only a handful were destined to become real independent producers. In the backward colonial society of the times the richer, well established merchants soon controlled the better part of the cigar production and the market.

A process of natural selection followed and from the multitude of petty producers who strived to conquer the promising cigar market, only the fittest, strongest and most skillful survived and moved on to find a solid place in the new and vigorous industry. Forty years after the race took off, the majority of the factories that would outlive the 19th century were already firmly established. Today, at the closing of the 20th century, many of those brands are still unsurpassed in the world of Havana cigars.

Competition, and the need to satisfy the tastes and requirements of the growing market forced the producers to constantly improve the selection and handling of the raw material, the mixing and blending of different kinds and strains of tobacco, the techniques used for manually making cigars, their packaging and marketing. By the year 1860, the big cigar producers in Havana had established their exquisite and finely packaged vitolas that remain mostly unchanged to this day. The gum used for modeling the drawing end of cigars was upgraded from a dilution of flour, to starch, to bread dough and finally to tragacanth, which is still in use. The gauge, weight and length of cigars gained uniformity and the production became standardized to near perfection nationwide.

In view of the constant orders that came in from Spain, including the Royal House, and from other foreign markets, cigar manufacturers developed special makes to please specific clients. Innovative vitolas of the most diverse gauges and sizes expanded the highly specialized craft of the cigar maker.

Ramon Allones, owner of the **La Eminencia** factory and the cigar brand **El Designio**, was the first to introduce the superior vitolas or *regalias* and in 1859 the luxury containers that captivated princes, kings and emperors alike, forcing all producers to follow suit in order to keep in pace with their clients demands.

The ungainly leather bales, barrels and trunks gave way to finely tooled cedar boxes and other luxurious containers initially designed and developed by Allones. Thus began the most refined, sophisticated and exciting form of publicity ever for a Cuban product.

The new containers were adapted to the existing varieties of vitolas allowing for a dis-

159

**E**stos MONTECRISTO de tamaño mediano y pequeño (sobre todo el N° 4, que es el HABANO más vendido del mundo), te permitirán vivir la leyenda en cualquier momento; quizá en algún intermedio que tengas en mitad del trabajo cotidiano. También puedes aprovechar para llenar de magia el tiempo muerto de alguna espera. Se lo pasan muy bien con tus amigos en las charlas de café y saben disfrutar como nadie de los buenos partidos. Pero lo que mejor se les da, y se quejan porque tienen pocas oportunidades de hacerlo, es pasear contigo al caer la tarde.

**Modern shapes (except for No.2, the pyramid or torpedo) are easier to make than most of the previous vitolas (shapes, sizes). Illustration courtesy of Habanos.**

B    JOYITAS    N°.3    N°.4    N°.5

Los MONTECRISTO de estas dimensiones se merecen ya que les des un poco más de protagonismo. Les gusta ser los reyes de las sobremesas porque se sienten importantes después de una buena comida. Ponen el detalle de lujo en cualquier celebración y tenemos noticia de que suelen crecerse mucho se les hace el honor de sellar con ellos un buen negocio. Pero nos consta que también saben hacerse íntimos cuando les das la confianza de acompañarte en casa, después de una larga jornada de trabajo.

Nº.1   Nº.2   ESPECIAL Nº.2   TUBOS   ESPECIAL

play of elegant and attractive boxes which the consumers soon learned to identify by the trademarks of their preference. Cedar became the standard wood for cigar box makers due to its exquisite fragrance that not only preserves but enhances the robust aroma of tobacco. The wooden parallelograms of different sizes contained the various kinds of vitolas carefully grouped by color and individually banded, because since 1861 each cigar manufactured for sale had to carry a paper ring by law (1). To this day, cigar boxes are wrapped and lined with 90 or 120 grams chrome paper with splendid full-color lithographies, some even in relief, called *habilitaciones*, the most striking of which appear on the lid, the underside of the lid or *vista*, and on the flap or *bofeton*.

The generalized use of small cedar boxes for the packaging of cigars in Cuba gave birth to a specialty in carpentry: box making. Likewise, the use of lithographies to wrap, adorn and seal the cigar boxes was the decisive factor in the introduction, development and profficiency attained by the lithographic industry in the country.

About that time the most noteworthy workshops, both by the number of workers as by the quality of their finished products, began to be called factories. Out of the 1217 cigar factories of all kinds that existed then in Cuba, 516 were found in Havana. These turned out the export quality cigars made with the best leaves, rolled by the most skillful craftsmen and sold at the highest prices.

In times of unusually great demand, the factories subcontracted their cigars with other workshops, but the raw materials always had to be first-class and supplied directly from the main factory. All other standards had to be strictly observed.

Many producers, in addition to their main brand, registered many others to flatter clients and honor celebrities of the day. For this purpose, the bands introduced in 1861 became veritable works of art when color lithography made it possible to reproduce the effigies of the rich and famous or of the buyers of a specific

1. The Havana skyline.

2. The Vuelto Abajo, Pinar del Rio, and its red soil that is unique in the world.

3. Seedlings being planted in soil that has been fertilized with natural materials.

vitola. The design, content and size of the bands was so varied that many beautiful and valuable collections of cigar bands are treasured today.

To obtain the necessary raw material for their factories many big manufacturers bought the most valuable tobacco plantations. Such was the case of Jaime Partagas, Antonio Rivero, Segundo Alvarez and later on, of Jose Gener, Prudencio Rabell, Manuel Rodriguez and others.

Necessity brought in specialization in each of the operations performed in the cigar-making process: storing, fermenting, loosening, moisturing, stripping, sorting, grading, blending, weighing, rolling, classifying, banding, wrapping, filleting and packaging which gave rise to the corresponding workers guilds and associations.

The problems and issues that the industry had to confront in its evolution affected the factory owners and the salaried workers alike. Class-based organizations were formed to defend the respective interests.

The determining factor in the evolution of the industrial process of tobacco and the advancement and solid training of the workers and managers that kept the factories running smoothly was and still is practical experience. The system of apprenticeships introduced in the 1830s, was extremely hard and exacting. Even the relatives of the producers who wanted to learn the trade had to slowly climb the ladder. A case in point was Jose *Pepin* Rodriguez, who in 1866 was sent at age 9 from Asturias to Cuba to learn all about the tobacco trade with one of his uncles in Havana. He did every single task that can be done in a factory, which gave him a solid qualification and a vast experience. He eventually became the sales manager of a famous brand and finally the proprietor of his own factory.

The procedures applied to tobacco from the moment it enters the factory as raw material until it comes out transfromed into a cigar, has remained unchanged for more than 150 years. Tobacco leaves reach the factory packed in *tercios* (2) or bales, while finder and filler leaves come in bales, where depending on their classes - they undergo a new period of fermentation which is where fermentation (3),

**1. A typical tobacco farm in Pinar del Rio with the drying sheds in the background.**

**2. Tobacco dries and comes in many natural shades, colors and tastes.**

**3. Tobacco is fermented, aged and shipped in special natural wrapping.**

the last phase of its aging (4), begins. The leaves began to age while hanging in the *cujes* or boughs at the plantation's curing barns and then in the pilon (5) or pile.

Tobacco bales are carefully stacked in the factory's warehouse and their position is periodically exchanged so that no same bale is placed twice under the weight of the rest of the stack. Bales containing special quality tobacco are placed in such a way that the carrots (four hands put together) lay horizontally, to stimulate fermentation, while bales containing light or over-dry tobacco are placed with the sheaves hanging vertically. The aging process is completed there at the factory warehouse, where the bales may remain for some time, depending on the quality of the leaves.

There are two methods for making cigars

**Tobacco buyers come from all over the world to examine the capa (wrapper) produced in Pinar del Rio. Large cigars can only be made with large leaves. These buyers are from Paris, France.**

manually. The *bunch* method consists of pressing filler leaves wrapped in binder between wooden boards hollowed lengthwise to serve as molds. After pressing for 20 to 30 minutes, the cigar maker rolls them over to eliminate any possible edges and then finishes the bunches or *zorullos* with wrapper leaf. The export quality vitolas or *regalias* are entirely hand-made by experts who give their highly personal touch to each cigar, turning them into finve Havanas.

At the factories, tobacco leaves have to be handled with the utmost care. The bales containing the wrapper leaves that are to be processed are opened in the storage room and the *gavillas* (6) or sheaves are taken to the area where loosening and moistening are performed. The leaves are loosened one by one from the sheaves and the following day they are thinly sprayed with water. In the old days the leaves were sprayed with a solution called *betun*, a sort of lye made with tobacco veins collected during the stripping phase and left to macerate in water. Some manifacturers mixed tobacco veins, honey and *"Aguardiente"* - the sugar cane distillate - to make their own special *"betun"*. The moist leaves are left to air for another 24 hours and then taken to the stripping department where the central vein of each leaf is removed totally, if the leaf qualifies as wrapper, or partially if it is earmarked as filler. Wrapper leaves are separated into right and left sides, for each of these have to be rolled accordingly. The stripped leaves are placed on *parrillas* (7) or shelves to loose excess humidity and then put away in barrels. The period of time they are to remain on the shelves and in the barrels depends on the weather conditions that prevailed during the harvest and the relative humidity in the atmosphere. This phase is closely controlled by the technicians to determine the exact moment when the leaves are ready for the cigar makers bench.

But before they reach the rollers, leaves have to be classified by the *rezagador* (8), an expert in charge of selecting and sorting out wrapper leaves from filler. Wrapper leaves are usually grouped by size into small, medium, large and giant; according to their color they can be light, straw-clear, vital-clear, light red, red, dark red and mature. Their texture can be fine or coarse and their (9) overall quality is considered good when the leaf is unbroken, spotless, uniformly colored and its veins are soft and subtle, and bad when the leaf is broken, stained, shows different colors and its veins are whitish and prominent.

The leaves classified as filler (10) and binder (10) are blended according to the requirements of a specific vitola (12) and also following the

**Cuba's best kept secret is this small machine about 15 inches wide and 24 inches long. It can make 400 cigars a day. The cut tobacco is put into the deck (above), then a leaf as wrapper is placed on the space in front of it. The handle in the rear is pulled, and out comes a cigar which just needs the finishing touches.**

tastes prevailing in the market. Blending is an all-important operation if a Havana is to achieve its exact flavor and aroma. To make a proper blend a skilled technician needs large stocks of tobacco leaves of different origins, ages and harvests.

1. Eumelio Espino Morrero, a scientist with the Ministry of Agriculture, invites the camera to visit his laboratory in Havana province where excellent cigars are made and grown.
2. Inside Eumelio's laboratory, shade grown plants are grown. Eumelio is responsible for the development of strains which produce most of Cuban cigar tobacco.
3. Stripping the central vein from the capa leaf.
4. Gathering the two halves of the leaf for storage and further fermentation.

Maintaining a blend invariable through the years is a manufacturer's permanent concern and a credit to his craftsmanship.

Filler leaves are generally classified by their strength into *volado*/strength 1, dry/strength 2, and light/strength 3. The "volado" leaves contribute aroma and combustibility, the "dry" determine taste and the "light" introduce strength. According to their size they can be large/no.15, medium/no.16, or small/no. 17. Normally filler leaves are blended to balance their strength. For instance, thin cigars with ring gauges 25 to 28 usually take only no. 1 strength filler; ring gauges 36 to 38 take a blend of nos. 1 and 2, while cigars with ring gauges over 38 take a combination of nos. 1, 2, and 3. Once the blend is done, the tobacco leaves are sprayed lightly with water and placed in large, closed wooden chests called *blending boxes*. These operations allow the tobacco to reach its peak condition.

Before turning over the raw material to the cigar rollers, each type of tobacco blend is carefully weighed in the metric scale. Each bundle of 50 cigars of a given vitola must have a fixed weight; since raw tobacco is bought by weight, once it is turned into finished cigars it is also sold by weight. That explains the standardized length and gauge for each vitola and the fixed weight for every 50 identical cigars. Weighing the raw material also helps to keep record of how much tobacco a cigar maker consumes per day.

The batches of mixed filler leaves, wrapped in moist cloth, are prepared in fixed quantities to make 75, 80, 90 and 100 cigars. Tobacco ready for the bench must contain no more than 14 to 15 per cent humidity, that is why it is kept in crates covered with damp burlap blankets in the galleys.

The workshop where the cigar makers roll cigars at their individual benches is called the galley and occupies the largest area in the factory. On his bench, a cigar roller has a square board of hardwood about two inches thick, a half-moon shaped cutting instrument called *chaveta*, a small guillotine, a wooden cigar pattern and a basin with tragacanth. The type of pattern he receives at the beginning of the shift indicates the kind of vitola that he will be making.

The cigar maker first takes a binder leaf, and with both hands spreads it totally flat on the board. Then he takes filler leaves with his right hand and rolls them with his left to make a uniform cylinder or *zorullo* properly compressed to allow the air to flow smoothly through it. He rounds off the drawing end eliminating any particles that might obstruct the careful finish that the tip requires. He places the *zorullo* over the

1. The aged, fermented capa halves are sorted by color.

2. The roller selects the leaves needed to make the type of cigar on which she is working.

3. Many rollers work together, but each makes the cigar completely by herself. These views are at the Partagas factory in 1995.

1. The cigar is rolled, a type of glue is applied to the end of the cigar and a cap is fastened to the end of the cigar.
2. Special knives are used by rollers. The blades have no handles.
3. The cigars are sorted by color.
4. The rolled cigars are kept in a mold for about half an hour or more.
5. A view into half a mold.
6. The cigars are tied in bundles of 50 after having been sorted by color. The, depending upon the need, they are ringed (banded), packed and shipped under almost any label in which the size is standard.

leaf on the board and with a rolling motion wraps the entire length of the cigar. New and precise cuts are needed to form the tip.

Holding the cigar with his left hand, he applies a minimal amount of gum on the little flap of wrapper leaf that forms the tip and gives the finishing strokes with his right thumb and index finger. Then he cleanly cuts the open end of the cigar with the guillotine or the *chaveta* and checks the ring gauge and length of the finished cigar with the pattern.

During his shift, a cigar maker repeats these operations as many times as his skill and dexterity allow. At the end of the day he ties his production in bundles of 50 cigars called *medias ruedas*. These ties are lightly fumigated and left to dry for at least at least two weeks in cedar chests at a temperature of 16 to 18 degrees Centigrade and 60 to 65 % humidity.

Once dry, the cigars are taken to the sorting room where the selector classifies, grades and groups them together in boxes. He puts the greatest care in making harmonious layers of cigars that perfectly match in color and shape. This procedure is done so scrupulously that each box shows almost absolute uniformity.

To sort out cigars by color, the selector divides them into 14 classes, of which 7 are dry and 7 spotted. Each of these classes are subdivided into six basic color groups: *"encendido"*, *"claro encendido"*, *"colorado"*, *"colorado pajizo"*, *"pajizo verde"* and *"verde"*, each of these ranging from the lightest to the darkest hues. This peculiar division of colors implies that the selector must normally display at least 70 groups of cigars on his table, and occasionally even more than a hundred.

Cigar selectors are highly qualified specialists who get paid accordingly. Their training takes at least five years of almost daily practice, provided they have the indispensable ability to distinguish an impressive range of colors and hues.

Once classified and sorted, the cigars are taken in their boxes to the banding department to be individually banded and packed in *"petacas"* (13), aluminum tubes, wrapped in cellophane paper or simply laid carefully in three types of cedar boxes: standard, natural door lid semi-boite (SBN-D), natural sliding lid semi-boite (SLB) and natural or varnished 8-9-8. The banding operators arrange the cigars in their boxes in exactly the same layer-patterns established by the selector.

The standard box is made of cedar and lined with 90 or 120 gram chrome paper with colourful lithographies called *"habilitaciones"*. There are seven different kinds: the *"cubierta"* or cover, *"vista"*, *"bofeton"*, *"papeleta"*, *"tapaclavo"*, *"largueros, y costeros"* and *"filetes"*. The *"cubierta"* covers the outer side of the lid, while the *"vista"* goes on the inner side. The *"bofetón"* is the paper flap glued to the inner back of the box to cover the cigars; the *"papeleta"*, with the trade mark or emblem of the manufacturer, goes on one of the outer sides of the box; the *"largueros y costeros"*, glued on each side of the box, bear the name of the vitola and the number of cigars contained in each box. The *"tapaclavos"* is either round or oval shaped and covers the nail that fastens the front end of the lid to the box, while the *"filetes"* are the bands glued to the outer angles of the box.

Cigars are placed in layers inside the lined boxes together with a slip that reads "Genuine Havana Cigars. Look for the Seal of Warranty of the Republic of Cuba on the box" in Spanish, English. French and German. The lid is then closed and fastened with a small bronze nail. When the *"tapaclavo"* the *"filetes"* and the Seal of Warranty are glued on, the box is ready for the market.

The semi-boite container with clasp (SBN-B), is made of natural wood without any paper lining and all their markings are fire-branded. The sliding lid on the identical (SLB) type opens back and forth horizontally. The 8-9-8 come in two types: natural or varnished. All semi-boites also carry the slip and the Seal of Warranty.

The finished boxes are packaged and taken to the fumigation chamber, sprayed and sent to the export warehouse. Pinewood crates are used for sea shipments, while cardboard cartons are used for air cargo. Both of these containers are marked and strapped for safety.

**The finished cigars are measured for uniformity of length and ring gauge.**

169

1-**Band**: A strip of printed paper that encircles each cigar usually near one of its ends and shows the manufacturers name or brand.

2-**Tercio**: A bale containing 80 bunches of classified tobacco leaves, wrapped with palm bark and tied with ropes or strings of majagua wood. It is used for transporting and storing tobacco.

3-**Fermentation**: The heating process by which tobacco loses excess resins and nitrogen compounds while aquiring uniformity in color, taste and aroma.

4-**Aging**: A process that begins when the tobacco leaf is cut and ends when it is ready for the cigar-maker's bench.

5-**Pilon**: Heap or pile of sufficiently dried tobacco leaves in the curing barn. Their dimensions vary according to the type of leaf. They must always be covered on the sides and top for protection.

6-**Gavilla**: Selected leaves that must fit into the left hand of the "engavillador" and are tied together by the thickest end of the central veins. The gavillas are made with a single class of leaf, wrapper or filler, always of the same age. Four gavillas make a bunch.

7-**Parrillas**: Wooden shelves on which the stripped tobacco leaves are laid to eliminate excess moisture.

8-**Rezagador**: The expert in charge of selecting wrapper, binder and filler leaves at the factory. A rezagador needs about five years of training before he is able to sort and select some four thousand wrappers a day.

9-**Wrappers**: The wrappers also are classified into clear, strawlike, mature, brown and ox blood.

10-**Filler**: The rolled tobacco leaves that form the core of a cigar. The filler is covered by the binder and then by the wrapper leaves. Also the branch whose leaves cannot be used as either binder or wrapper.

11-**Binder**: The leaves used as first cover of the filler.

12-**Vitola**: Size and shape of rolled cigars with a specific length, ring gauge and weight. Their names vary according to the producer, though Coronas, Medias Coronas, Petit Cetros, Nacionales, Brevas and others are almost identical regardless of the brand. Some manufacturers boast more than 300 different vitolas in their inventories, among them Heraldos, Monarcas, Palmas, Victorias, Caramelos, Churchills, Amatistas, Belvederes, Miniaturas, Perfectos, Panetelas, Macanudos, etc.

13-**Petaca**: Pocket-size cigar containers usually made of cardboard or aluminum.

14-**The Seal of Warranty**: It is a green, paper sticker 6 3/4 inches long and 2 3/16 inches wide which was introduced by a Roayl Decree on 13 February 1889, to be placed on all cigar, cigarrete and pipe tobacco containers manufactured by the **Union de Fabricantes de Tabacos de La Habana** in order to certify the origin of these products and protect them aainst forgeries and disloyal competition against Cuban brands abroad.

Originally the seal bore the name of the **Union de Fabricantes de Tabacos de La Habana** on the top, the Spanish coat of arms on the left and the Seal of the Governor General on the right.

**This was the seal of warranty produced in 1899. A newer seal has taken its place**

When the Republic was innaugurated in 1902, the manufacturers union requested authorization from the Government of Tomás Estrada Palma to modify the seal in consistency with the new reality. The new design substituted the Spanish emblem for the Coat of Arms of the Republic of Cuba, and the seal of the former Governor General for an effigy of Christopher Columbus. This design was duly registered and used until the passing of the Valdés Carrero Act in 1912. On 16 July that year, the House of Representatives adopted a Bill introduced by Luís Valdés Carrero, who came from the cigar makers ranks, authorizing the President of the Republic to create a Seal of Warranty of National Origin, to oficially protect the Cuban cigar industry against forgeries and imitions of its products abroad.

The provisions of the Valdés Carrero Act were regulated on 9 October, 1912. The Seal of Warranty of National Origin was registered as a national collective brand of the Republic of Cuba at the Department of Agriculture, Commerce and Labor. Since then the seal is glued to all containers used for export to certify the quality and authenticity of the product.

In January 1931, the seal was redesigned into its present form with the Coat of Arms of the Republic on the left and a scene with harvesters working on a tobacco plantation on the right.

# INDEX

Alvarez, Lopez & Co., 22
A.R.Suaraz, 55
Artemisa, 11
Benito Celorio, 28
Berarda Salabarria, 4
Bock label, 54
Buenaventura, 11
Caimito, 11
Calixto Lopez, 32
Carlos Manuel de Cespedes, 33
Cayo La Rosa, 11
Cienfuegos, 11
Comision Nacional de Propaganda, 13
Compania Ll. de la Habana, 2
Cortes, 18
counterfeiters, 21
Cuban Land and Leaf Tobacco, 53
Desiderio Camacho, 16
Don Benito Celorio, 42
Don Francisco de Arango y Parreno, 17
Don Saturnino Martinez, 28
Don Segundo Alvarez, 29
Dr. Don Juan Bances Conde, 38
Dr.Jose Manuel Cortina, 74
Eden de Bances rings, 31
El Dinamico, 61
El Hurrican, 22
El Rey del Mundo label, 65
El Rey del Mundo ring, 36
Elios, 70
fakes, 21
Ferdinand VII, 18
Finca Ajiconal, 60
Flor de Alvarez, 57
Flor de Astor, 57
forgeries, 31
Francisco Perez del Rio, 62
frauds, 31
Garriga y Villa, 54
General Antonio Maceo, 47
Gonzalez del Valle, 22
Govea, 11
Gustavo Bock, 43
habano defined, 13
Havana cigar, 13
Henry Clay, 51
Hoyo de Monterrey, 48
H.Upmann, 23
Jaime Partagas, 22
Jose Alonso, 58
Jose Gener, 34
Jose Jimenez Perez, 2
Jose Marti, 47
J.Suarez Murias factory, y, de, 25
Julian Rivero, 37
La Carolina, 58
La Corona, 22
La Corona rings, 30
La Diana, 64
La Intimidad, 51
La Meridiana, 22
La Preferencia, 73
La Radiante, 52
Las Tunas, 7
Lopez Hermanos, 33
Lord Byron, 75
Lord Lonsdale, 50
Manuel Alvarez, 56
Map of Cuba, 8
Mara Duharte, 4
Maria Guerrero, 39
Murias rings, 29
Nancy Barreras, 7
Paletas, 11
Partagas rings, 24, 27
Partido, 11
Perez y Diaz label, 59
Piedras, 11
Pinar del Rio, 11
Por Larranaga, 22, 69
postage stamps, 10
Pres. William McKinley, 41
Punch, 4
Real Sociedad Economica, 20
Romeo y Julieta rings, 12, 30
Sancti Spiritus, 11
Santa Clara, 16
Santa Damiana label, 63
Santa Felipa, 55
Secota Village, 14
Semi Vuelta, 11
Serbio, 61
Tumbadero, 11
vanity bands, 23
vanity rings, 24, 27
vegueros, 11
Villa Clara, 11
Vitofilia Habana, 4
Vuelto Abajo, 11

# BIBLIOGRAPHY

**Alfonso San Juan, Mario**: EL TABACO. Editorial "Mesquita."

**Archivo Nacional de Cuba**: Brands and Trademarks Collection. Files 198-200-206-201-202-203-204-205-207-208-209-210-211-212-213-214-256-244-197-253-252-251-250-249-248-247-246-245-254-255. Books 1-8 of the Clerkships Collection. The Jose C. Barrena Archive, File 141, No. 1, years 1894-1895. Clerkship of Valerio Ramirez, File 391, years 1885-1886.

**State Provincial Archive of Pinar del Rio**: Colonial Judicial Institutions. Mayoralty of Pinar del Rio. Dossiers 3380, 2435, 2579, 2610, 2638, 2684, 2728, 2745, 2762, 2761, 2913, 3028, 3052, 3073, 3099, 3100, 3155, 3159, 3206, 3207, 3209, 3210, 3312, 4116, 4207, 4372, 4373, 4437, 4739, 5086, 3317 in connection with Jaime Partagas Ravelo.

**Arredondo, Alberto**: LUCHA DE EMPRESAS. INDUSTRIALISMO. Revista Tabaco, March-April, Havana, 1944.

**Asencio, Manuel**: REFLEXIONES Y CALCULOS SOBRE EL DESESTANCO DEL TABACO. Imprenta del Gobierno y Capitania General por SM. 2nd edition. Havana, 1879.

**Bacardi Moreau, Emilio**: CRONICAS DE SANTIAGO DE CUBA.

**Casado, Ricardo A.**: NUESTRO TABACO, EL HABANO SIN IGUAL. Havana, 1939.

**Celorio Hano, Benito**: CARTA AL GOBERNADOR GENERAL DE LA ISLA DE CUBA. January 27, 1879.

**Collado, Gonzalo**: SERAN LOS TORDEDORES LOS QUE DIRAN LA ULTIMA PALABRA SOBRE EL MAQUINISMO. In "El Tabacalero," October. Havana, 1947.

**Comision para el Estudio del Maquinismo en la Industria del Tabaco**: INFORME. In "El Tabacalero," November, 1945 and August and December, 1946.

**Comision Nacional de Propaganda y Defensa del Tabaco Habano**: MEMORIAS. Years 1930, 1931-32, 1934, 1936, 1937, 1938, 1940-41, 1942, 1943, 1944, 1945, 1946-1950, 1951, 1952, 1953, 1054, 1955, 1956, 1957, 1958.

**Cortina, Humberto**: TABACO, HISTORIA Y PSICOLOGIA. La Habana. Imprenta P. Fernandez y Cia. Havana, 1939.

**Cubatabaco**: Showcase of the different brands of cigars and cigarrettes manufactured in Cuba. Album. Havana, 1967.

**Cuba**: Laws, Decrees et al. POLITICA TABACALERA DEL GOBIERNO, bajo la orientacion del Dr. Eduardo Suarez Rivas. Edit. P. Fernandez y Cia. Havana, 1955.

**De Juan, Adelaida**: PINTURAS Y GRABADOS COLONIALES CUBANOS. Havana.

**De la Sagra, Ramon**: CUBA 1860. Comision Nacional Cubana de la UNESCO. Selected articles on Cuban agriculture. Havana, 1963.

**De Mas y Otzet, Francisco**: EL TABACO Y LA INDUSTRIA TABACALERA EN CUBA. La Habana, 1886. "Jose Marti" National Library.

**Diaz Irizar, Mario**: MARCAS Y PATENTES. La Habana, 1917. "Jose Marti" National Library.

**Diccionario Enciclopedico Hispanoamericano**: EL TABACO. Vol. XXI. C.H. Simonds Company. Boston. USA. Undated.

**Felipe, Victoriano**: EL TABACO. Imprenta F. Furtanet. Madrid, Espana, 1851.

**Fernandez de Madrid, Jose**: MEMORIA SOBRE EL COMERCIO, CULTIVO Y ELABORACION DEL TABACO EN ESTA SIEMPRE FIEL ISLA DE CUBA. La Habana, 1822.

**Fernandez Roque, Mario**: CUBA, EL PAIS DEL TABACO HABANO, in Libro de Cuba. La Habana, 1953. "Jose Marti" National Library.

**Friedlander, N. E.**: HISTORIA ECONOMICA DE CUBA. Ed. J.Montero, La Habana, 1944.

**Garcia Galan, Gabriel**: EL TABACO Y SU ACCION EN LA INDEPENDENCIA DE CUBA. La Haban, 1958.

**Garcia Gallo, Gaspar Jorge**: BIOGRAFIA DEL TABACO HABANO. La Habana, 1959.

**Garcia Marques, Rafael**: INFORME AL EXCELENTISIMO SR. M.H.MACKINLEY, PRESIDENTE DE LOS ESTADOS UNIDOS DE AMERICA, ACERCA DE LA GRAVE SITUACION EN QUE SE ENCUENTRAN LAS INDUSTRIAS TABACALERAS, CAUSAS DE SU DECADENCIA Y MEDIDAS QUE SE CONSIDERAN NECESARIAS PARA SALVARLAS DE LA RUINA QUE LAS AMENAZA. Union de Fabricantes de Tabacos y Cigarros de La Habana. Havana, March 27, 1900.

**Garcia de Torres, Juan**: EL TABACO. CONSIDERACIONES SOBRE EL PASADO, PRESENTE Y PORVENIR DE ESTA RENTA. Imprenta de J. Moguera a cargo de M. Martinez. Madrid, Espana, 1875.

**Gener Batet, Jose**: PROYECTO PARA RESOLVER LA GRAVE CUESTION ECONOMICA DE LA ISLA DE CUBA, in Vidal Morales y Morales. Factual Collection, Vol. 61, No 6. La Habana, 1873.

**Gonzalez Aguirre, Jose**: LA VERDAD SOBRE LA INDUSTRIA DEL TABACO HABANO. imprenta de P. Fernandez y Cia. La Habana, 1905.

**Gonzalez del Valle, Angel**: MEMORANDUM PRESENTADO A LA COMISION NACIONAL DE PROPAGANDA Y DEFENSA DEL TABACO HABANO. Memoria. La Habana, 1929. "Jose Marti" National Library.

**Gomez Flores, Emilio**: EL TABACO. Tipografia de Manuel G. Hernandez. Madrid, 1889.

**Gordon, Antonio**: EL TABACO EN CUBA. APUNTES PARA SU HISTORIA. Obsequio de la Real Fabrica de Cigarrillos y Picaduras "La Legitimidad," de Prudencio Rabell. La Habana, 1897. "Jose Marti" National Library.

**Gran Hotel Pasaje**: GUIA Y RECUERDO (tourists brochure). La Habana, 1890.

**Guevara de la Serna, Ernesto "Che"**: OBRAS 1957-1967. Edit. Casa de las Americas. Vol 1. La Habana, 1970.

**Hazard, Samuel**: CUBA A PLUMA Y LAPIZ. Ed. Cultura S.A. La Habana, 1928.

**Interviews and testimonies**: German Enrique Upmann Machin, Crisanto Cardenas, Martha Cifuentes, Joaquin Gomez, Julio Le Riverand, Eduardo Rivera, Emilia M. Tamayo, Lazaro Garcia, Magali Garcia, Humbarto Cabeza, Orlando Arteaga.

**Jimenez, Juan Bautista**: UNA ESCOGIDA DE TABACO. (booklet). La Habana. Undated.

**Libro de Cuba**: Volumes 1902-1952. La Habana. "Jose Marti" National Library.

**Lugo de la Cruz, Evelio**: LOS COSECHEROS Y LA MECANIZACION DE LA INDUSTRIA. In "El Tabacalero," 1,8 December. Havana, 1945.

—LOS FABRICANTES Y LA MECANIZACION. In "El

Tabacalero," 1,8 January. Hav.1946.

—LA REUNION DE AGRICULTURA Y LAS MAQUINAS DE HACER TABACOS. In "El Tabacalero," December. Havana, 1947.

**Llaguno y de Cardenas, Pablo**: EL TABACO. ESTUDIO SOBRE EL CULTIVO DEL HABANO Y SUS SUPLANTACIONES, CNPDTH. La haban, 1945. "Jose Marti" National Library.

**Magazines**: HABANO. Organo Oficial de la Asociacion de Almacenistas y Cosecheros de Tabaco de Cuba. Editorial Habano S.A. Revista TABACO. Organo Oficial de la Federacion Abacalera Nacional. La Habana. 1933, 1934 and 1945.

**Marrero, Gregorio**: LA INDUSTRIA TABACALERA EN CUBA Y LA MAQUINARIA INDUSTRIAL. Federacion de Torcedores de Cuba. La Habana, 1927.

—LA MECANIZACION EN LAS FABRICAS DE TABACO. In "El Tabacalero" 14 June. Havana, 1946.

**Nunez Jimenez, Antonio**: EL VIAJE DEL HABANO. Empresa Cubana del Tabaco, La Habana, 1988.

—CUBA EN LAS MARQUILLAS DEL SIGLO XIX.

**Ortiz, Fernando**: CONTRAPUNTEO CUBANO DEL AZUCAR Y EL TABACO. Consejo Nacional de Cultura, La Habana, 1963.

**Pardo y Betancourt, Valentin**: INFORME ILUSTRADO Y ESTADISTICO SOBRE LOS ELEMENTOS DE RIQUEZA DEL TABACO EN EL Ano 1861. La Habana, 1863. "Jose Marti" National Library.

**Parraga y Mendez Capote (Law firm)**: ESCITO DE REPLICA PRESENTADO A NOMBRE DE LA HAVANA COMMERCIAL COMPANY. La Habana. Undated.

**Perdomo, Jose E. y Jorge J. Posse**: MECANIZACION DE LA INDUSTRIA TABACALERA. La Habana, 1945.

**Perdomo Rivadeneira, Jose Enrique**: EL COMERCIO TABACALERO CUBANO. Editorial Habano. La Habana, 1949.

—TABACO HABANO. SITUACION ACTUAL DE SUS MERCADOS EXTERIORES Y POSIBILIDADES DE EXPANSIONAR SU COMERCIO. Imprenta El SigloXX. La Habana, 1945.

—LEXICO TABACALERO CUBANO. 1ra Edicion. La Habana 1940.

**Porto Capote, Gustavo A.**: NUESTRO TABACO EN LA POST-GUERRA. La Habana, 1945.

**Registro Mercantil de La Habana**: Libro de Sociedades. Vols. 362, 128, 294, 599, 74, 127, 241.

**Registro de la Propiedad de San Juan y Martinez**: Vols. 3, 6, 7, U7, 9, U10, 12, 13, 15, 17, 19, 22, 24, 28, 29, 30, 31, 32 and U42; files in connection with Jose Gener Batet, his brothers and descendants; Cifuentes y Compania; Francisco Pego Pita; Cifuentes, Fernandez y Cia; Cifuentes, Pego y Cia.

**Reinoso, Alvaro**: DOCUMENTOS RELATIVOS AL CULTIVO DEL TABACO. Edit. La Propaganda Literaria. La Habana, 1888. "Jose Marti" National Library.

**Rivero Muniz, Jose**: TABACO. SU HISTORIA EN CUBA. Instituto de Historia. Comision Nacional de la Academia de Ciencias. La Habana, 1964-1965. 2 volumes.

—TABACO; ENSAYO DE UNA BIBLIOGRAFIA TABACALERA EN ESPANOL. Edit. P.Fernandez y Cia. La Habana, 1957. (As reprinted by the Revista de la Biblioteca Nacional, vol 2, no 1.)

—LA LECTURA EN LAS TABAQUERIAS. Edit. P. Fernandez y Cia. La Habana, 1957. (As reprinted by the Revista de la Biblioteca Nacional, vol 2, no 4.)

—EL MOVIMIENTO LABORAL CUBANO DURANTE EL PERIODO 1906-1911. APUNTES PARA LA HISTORIA DEL PROLETARIADO EN CUBA. Universidad Central de Las Villas. Santa Clara, Cuba, 1962.

—LAS TRES SEDICIONES DE LOS VEGUEROS EN EL SIGLO XVIII. Lecture delivered by the academician. Edited under the auspices of the Academia Nacional de Cosecheros de Tabaco de Cuba. La Habana, 1951.

**Rodriguez Ferrer, Miguel**: EL TABACO HABANO, SU HISTORIA, SU CULTIVO, SUS VICISITUDES, SUS MAS AFAMADAS VEGAS. Imprenta del Colegio Nacional de Sordo-mudos. Madrid, Espana. 1851. "Jose Marti" National Library.

**Rodriguez Lopez, Manuel**: INFORME SOBRE EL TABACO HABANO Y REMEDIO PARA SU CRISIS. Presented to the Comision Nacional de Propaganda y defensa del Tabaco Habano. La Habana. Imprenta Perez Sierra,1937.

**Rodriguez Navas, Manuel**: EL TABACO, SU CULTIVO, PRODUCCION Y COMERCIO. Edit. de Bailly-Bailliere. Madrid, Espana. 1905.

**Rodriguez Ramos, Manuel**: SIEMBRA, FABRICACION E HISTORIA DEL TABACO; CON EL MANUEL DEL TABAQUERO. La Habana, 1905.

**Roig, Juan Tomas**: DICCIONARIO BOTANICO DE NOMBRES VULGARES CUBANOS. La Habana. "Jose Marti" National Library.

**Samalea, Armando**: LA MECANIZACION DE LA INDUSTRIA. In "El Tabacalero," 4-5. La Habana, 1945.

**Sanchez de Fuente, Fernando**: PONENCIA PRESENTADA A LA COMISION ESPECIAL DE LA CAMARA DE REPRESENTANTES QUE ESTUDIA LA CRISIS DE LA INDUSTRIA TABACALERA Y EL PROBLEMA OBRERO. La Habana, 1916. "Jose Marti" National Library.

**Sociedad de Estudios Economicos**: EXPOSICION DIRIGIDA AL EXCMO. SR. MINISTRO DE ULTRAMAR ACERCA DE LA IMPORTANCIA DEL TABACO EN CUBA. Imprenta y Papeleria "La Universal," de Ruiz y Hno. La Habana, 1891.

**Stubb, Jean**: TABACO EN LA PERIFERIA. editorial Ciencias Sociales. La Habana, 1989.

**Trelles, Carlos**: MEMORIA SOBRE EL AZUCAR Y EL TABACO EN LA EXPOSICION UNIVERSAL DE PARIS DE 1900, presented to the Cuban Commissioner Sr, Gonzalo de Quesada. Edit. "El Figaro." La Habana, 1901. "Jose Marti" National Library.

**Union de Fabricantes de Tabacos de La Habana**: INFORME SOBRE VALORACION DEL NUEVO ARANCEL, APROBADO POR UNANIMIDAD EN JUNTA GENERAL EL DIA 5 DE MAYO DE 1893, DIRIGIDO AL MINISTRO DE ULTRAMAR. La Habana, 1893. "Jose Marti" National Library.

—MEMORIA DE LOS TRABAJOS MAS IMPORTANTES REALIZADOS POR LA CORPORACION DESDE EL 18 DE SEPTIEMBRE DE 1890 HASTA EL 5 DE FEBRERO DE 1894, EN DEFENSA DE LOS INTERESES GENERALES DE LA INDUSTRIA QUE REPRESENTA. La Habana,1894.

**United States Department of Agriculture**: TOBACCO IN THE UNITED STATES; PRODUCTION, MARKETING, MANUFACTURING, EXPORTS. Agricultural Marketing Service. Washington D.C. 1961.

**Vidaurreta Casanova, Antonio Julio**: EL TABACO EN CUBA; BREVE ESTUDIO HISTORICO-POLITICO-SOCIAL LITERARIO DEL TABACO. Edit. Culturales Publicidad. Santa Clara, Cuba. 1943.

**Vilardebo y Monet, Jose**: EL TABACO Y EL CAFE. SU HISTORIA, SU ACCION PSICOLOGICA Y SUS PROPIEDADES MEDICINALES. La Habana, 1860. "Jose Marti" National Library.

GRAN FABRICA DE TABACOS

FINEST PRODUCT OF THE ISLAND OF CUBA

DE SEGUNDO